A Handbook for Analytical Writing

Keys to Strategic Thinking

Synthesis Lectures on Professionalism and Career Advancement for Scientists and Engineers

Editors

Charles X Ling, University of Western Ontario
Qiang Yang, Hong Kong University of Science and Technology

A Handbook for Analytical Writing: Keys to Strategic Thinking
William E. Winner
2013
A Practical Guide to Gender Diversity for CS Professors
Diana Franklin
2013

A Handbook for Analytical Writing: Keys to Strategic Thinking
William E. Winner

www.morganclaypool.com

ISBN: 9781627051828 print
ISBN: 9781627051835 ebook

DOI 10.2200/S00484ED1V01Y201303PRO001

A Publication in the Morgan & Claypool Publishers series
SYNTHESIS LECTURES ON PROFESSIONALISM AND CAREER ADVANCEMENT FOR SCIENTISTS AND ENGINEERS
Lecture #1
Series Editor: Charles X Ling, University of Western Ontario, and Qiang Yang, Hong Kong University of
    Science and Technology

Series ISSN Pending

A Handbook for Analytical Writing

Keys to Strategic Thinking

William E. Winner
Department of Forestry and Environmental Resources, North Carolina State University

SYNTHESIS LECTURES ON PROFESSIONALISM AND CAREER ADVANCEMENT FOR SCIENTISTS AND ENGINEERS #1

MORGAN & CLAYPOOL PUBLISHERS

ABSTRACT

This handbook accelerates the development of analytical writing skills for high school students, students in higher education, and working professionals in a broad range of careers. This handbook builds on the idea that writing clarifies thought, and that through analytical writing comes improved insight and understanding for making decisions about innovation necessary for socioeconomic development. This short handbook is a simple, comprehensive guide that shows differences between descriptive writing and analytical writing, and how students and teachers work together during the process of discovery-based learning. This handbook provides nuts and bolts ideas for team projects, organizing writing, the process of writing, constructing tables, presenting figures, documenting reference lists, avoiding the barriers to clear writing, and outlines the importance of ethical issues and bias for writers. Finally, there are ideas for evaluating writing, and examples of classroom exercises for students and teachers.

KEYWORDS

writing, analytical, handbook, science, engineering, business, planning, management, decisions, STEM

Contents

Preface

The simple goal of this handbook is to accelerate the development of analytical writing skills for high school students, students in higher education, and working professionals in a broad range of careers. This handbook builds on the idea that writing clarifies thought, and that through analytical writing comes improved insight and understanding for making decisions about innovation necessary for socioeconomic development.

This handbook explains the differences between descriptive writing and analytical writing, and presents guidelines explaining how students and teachers work together to write and think during the process of discovery-based learning. This handbook provides nuts and bolts ideas for forming functional teams, organizing writing, the process of writing, and constructing tables, figures, and reference lists. This handbook provides tips to avoid the ten most common barriers to clear writing, and outlines the importance of ethical issues and bias for writers. Finally, there are ideas for evaluating writing and examples of classroom exercises for students and teachers.

This handbook fills an important need. Successful scientists, engineers, business people, and other professionals acquire basic writing skills in English courses in high school and college. But for many, such courses provide only the foundation for the writing they do in careers later in life. Most professionals begin learning their analytical writing skills in post-graduate education, with mentors who help students develop documents including theses and journal articles. Later, editors for books and journals provide more writing instruction, and students' writing skills mature out of necessity. Still later, young professionals critically read the writings of others, and provide critical reviews of reports, business plans, and grant proposals. With each review, writing skills improve.

More than a decade of teaching capstone courses to undergraduate students convinced me that students in high school and higher education can, and should, learn analytical writing skills. Teachers, instructors, and professors can use this handbook to teaching writing in a wide range of classes in disciplines ranging from Agriculture to Zoology. This handbook can be an important guide for students enrolled in project-based courses, capstone courses, or research-based courses for both undergraduate and graduate students. This handbook can also be useful for students preparing projects and theses for graduate degrees.

I wrote this handbook so that it does not require a teacher; any student or professional in a career can use this handbook to increase their analytical writing skills. This handbook is a self-teaching guide for anyone wanting to improve analytical writing and thinking skills.

William E. Winner
Department of Forestry and Environmental Resources
North Carolina State University

Acknowledgments

I wish to thank all those who encouraged me to create this handbook. My wife, Terri Lomax, is a constant source of support. My children, Ben and Maren, and my brothers and sister, Russ, Phil, and Carolyn, taught me life-long lessons about the importance of writing, oral communications, and working in teams. I sincerely appreciate the guidance from David Dalton and Hal Mooney.

Dedication

I dedicate this handbook to the loving memory of my parents, Jean and Lee.

CHAPTER 1

Introduction

1.1 WRITING ADVANCES THINKING

Students and teachers often overemphasize collecting information, while not placing enough importance on the analytical thinking and communication required for establishing discoveries. This handbook will help demystify analytical writing, and through writing will improve the quality of thinking and communication for ideas essential for making decisions in both the professional and personal arenas. The goal of this handbook is to facilitate writing skills that reduce barriers to completing projects, forming recommendations, and making decisions.

Analytical writing connects to critical thinking, as in the adage: "Writing clarifies thought." Writing forces more clear recognition of the issues because explaining ideas to readers reveals gaps in logic. Also, writing invariably leads to new ideas, even for experts who understand a field. Finally, writing lets others understand your thinking about a matter, attributes ideas to you, and becomes the basis for further discussion and writing. Writing advances thinking when authors create multiple drafts of manuscripts to improve organization, fill gaps in reasoning, or discover new issues that enrich understanding.

The analysis of information through writing and thinking is important because rational thinking is the foundation for making decisions in an increasingly complex world. In science, rational thinking is essential for coursework, term papers, project reports, and theses for undergraduate and graduate students. Analytical thinking allows professional scientists to synthesize information in ways leading to new ideas and discoveries. For engineers and those in business, analytical thinking is a guide for making decisions about resource investment, project evaluation and management, and projected outcomes of decisions. Outside the classroom, analytical thinking skills are essential for making decisions about which courses to take, how to vote, where to work, and other choices fundamental in life.

Analytical writing and thinking are more important today than ever before, and will increase in importance far into the future. Humans face increasing complex issues such as population growth, globally entwined economies, transformation to new technologies and policies for energy, providing health care, and mitigating and adapting to climate change. For example, large data sets of the future challenge our current abilities to transform the numbers into meaningful information. The field of "Analytics" is emerging to manage real time data for sales and inventory in retail shopping, the use of energy in smart grids and smart buildings, health care, and many other fields. Still, even with sets of large data, communication of analysis is at the heart of making decisions.

Analysis of information, including data and concepts, is the only approach for making strategic decisions for coping with a rapidly changing world. Students who learn analytical approaches to writing and thinking will find innovative opportunities to invent, manage resources and capital, and guide the human effort to optimize opportunities for all. In the absence of analytical, strategic approaches to making decisions, humans will be at a distinct disadvantage in managing the present or preparing for the future.

This handbook will help people become more proficient with the processes of analytical writing and thinking. Students in high school and in higher education serve themselves well by learning to write and think with analytical processes and tools. High school teachers, and instructors and professors in higher education, committed to preparing students for enriched lives must aid students with new opportunities to learn critical skills in communication and thinking. Those already in careers in both the public and private sectors can take steps to bring new, advanced ideas derived from analysis and strategic planning to their workplace. Those people, institutions, agencies, and businesses that advance analytical writing and thinking will find rewards, recognition, and successes along the way.

1.2   HISTORICAL EXAMPLES

History provides important lessons for those learning about analytical writing and thinking. Analytical thinking is certainly in our DNA, and led to survival of our species. But recognizing that analytical thinking requires analytical writing is recent in human history.

The process of analytical, scientific writing originates from the earliest experiments of Sir Francis Bacon (1561-1626) (Overton 1994). Bacon founded deductive reasoning by using information gathered through many sources, especially experimentation, to draw conclusions. Deductive reasoning replaced the Aristotelian approach to learning through observation. Bacon also wrote, publishing *Instauratio Magna*, *Novum Organum*, and *The Advancement of Learning*. The writings of Bacon established the scientific method, and the process of analytical writing as the requirement for progress and discovery.

Since Bacon, legions of scientists effectively used analytical, scientific writing to great advantage. The three historical examples that follow show how analytical writing played key roles in advancing science. The selected examples highlight situations that show the power of analytical writing, and how such writing leads to advances in the way people see the world and transform the way we live. The examples also show how students can avoid pitfalls in their efforts to tackle projects.

1.2.1   GREGOR MENDEL: DEVELOPING A BASE OF INFORMATION FOR ANALYSIS

A look at Gregor Mendel (1822–1884), famous for his work with pea plants, shows how collecting information led to a testable idea further developed by analytical thinking. Mendel wanted to know how to predict the outcome of breeding domesticated plants and animals. For thousands of years, people understood the wisdom of domesticating plants and animals through crossbreeding. Even though people had a basic understanding of inheritance, there was no capacity to predict the characteristics of offspring based on observing the parents. Mendel used analytical writing to ask an important question, gathered relevant information in written form so that patterns emerged from results, and communicated his ideas in formal writing to his peers. From his work comes our understanding of dominant and recessive genes, their role in inheritance, the concepts of segregation and independent segregation of genes, and ultimately the mechanisms of sexual recombination of genes from parents.

Mendel noticed two parent plants with different characteristics often produced seedlings similar to one parent, but sometimes different from either parent. Pea plant characteristics interesting to him included traits such as flower and pod color, floral anatomy, and physical characteristics of the stem. He eventually concluded that seven physical characteristics of flowers and stems occurred in only one of two forms. His selective breeding of over 28,000 plants, took many years, and through it all, he kept detailed notes on the seven characteristics of parent plants and resulting seeds and seedlings.

However, Mendel did much more than cross pea plants and record the resulting characteristics of seeds and seedlings. He used analytical thinking to seek a "law" that would explain inheritance of seedling attributes from parent plants. Mendel felt that by carefully recording the outcomes of experimental crosses and by forcing himself to write about the results, he could begin to predict the characteristics of seedlings based on parent traits. What was difficult and insightful was thinking through the patterns of physical traits in parents and offspring, and recognizing heritability for each quality from each parent. From the large tables of data, he could begin to see patterns and make predictions of the ratios of flower colors and plant height of seedlings based on characteristics of the parent plants.

Mendel further established the concept by taking the information from his own notebooks and formally writing documents that others could read, consider, evaluate, and test in their own ways. As was the convention in his time, Mendel presented his work in two lectures on February 8 and March 9, 1865, to the Natural History Society of Brno, in Slovakia. In 1885, Gregor Mendel presented his work at a scientific conference, and later published his work in a paper, *Experiments on Plant Hybridization*. He also published the lectures in an 1866 issue of *Verhandlungen des naturforschenden Vereins*, the *Proceedings of the Natural History Society in Brno* (Mendel 1866). Versions of his papers are now available in English (Mendel, republished, 2008).

1.2.2   JAMES WATSON AND FRANCIS CRICK: WORKING IN TEAMS AND ATTRIBUTION

Analytical writing often represents teamwork and writing. Unfortunately, confusion can exist when people, either working alone or on teams, do not acknowledge the importance of the work of others. In some cases, tensions between scientists and team members develop around the questions of listing authors of a paper, and if the level of effort required to be an autor on a document is sufficient. The tensions result when important projects carry prestige, or are connected to copyrights, patents, and professional advancement.

When in doubt, there is little penalty for including all those significantly engaged in a project as author, and inviting them to participate in writing. Each person listed as an author must play some role in both developing and writing the manuscript. Those listed as authors in early drafts of the manuscript can remove their own names from the title page if they feel inclusion is inappropriate. In general, problems that result from omitting a deserving author are larger than the problems from including a person who perhaps should not belong.

James Watson and Francis Crick provide a well-known example of the controversy that can emerge around issues of authorship. Of course, Watson and Crick are famous for discovering the double helix structure of DNA because they published their now famous paper in *Nature* (Watson and Crick 1953).

Watson and Crick formed their idea from critically thinking about data they had, as well as data from other sources. Watson and Crick, along with Maurice Wilkins, shared the 1962 Nobel Prize for discovering the DNA structure. These three scientists had a general idea about the structure of DNA, but were not quite able to visualize the arrangement of chemical bonds and the three-dimensional shape of the DNA molecule. Watson and Crick recognized the potential for X-Ray crystallography to reveal the physical structure of DNA molecules.

Watson and Crick worked at Cambridge University during the early 1950s when several other large research groups were also working to discover the structure of DNA. Wilkins and his colleague, Rosalind Franklin, were at King's College in London (Elkin 2003). Franklin used crystallographic techniques to photograph a DNA molecule, calculated angles and measurements of the double helix, and placed the image in her desk drawer. Without asking, Wilkins took the photograph from Franklin's drawer, and shared it with Watson and Crick who immediately set to work finalizing the chemistry that made the double helix structure of DNA.

The *Nature* article from Watson and Crick does acknowledge unpublished data from Wilkins and Franklin. However, many scientists feel Rosalind Franklin played such a critical role in the discovery of the double helix structure of DNA that she should have authorship in the *Nature* article. Franklin died in 1958 from cancer, perhaps from the X-ray technology she used, and therefore could not receive a Nobel Prize awarded in 1962. Today, many see her as a heroine of scientific research who did not receive the recognition she deserved, diminishing some of the luster from the fame Watson and Crick achieved.

The lessons from Watson and Crick go beyond the discovery of DNA. New concepts and discoveries almost certainly require information and data already in hand, and may also require data, ideas, and information coming from others. Analytical thinking requires building on information from others, as few discoveries are made in total isolation. Writing is one of the best ways to document how the foundation for the work developed by respectfully acknowledging the important contributions of others. Acknowledging sources of information and data used for further original thought is essential, never diminishes the importance of a discovery, adds to the network of those interested in the advance, and makes for a good citizen of science. So, although Franklin published scientific articles of her own, her exclusion from authorship on Watson and Crick's *Nature* paper, founded in large part on her work, demonstrates the importance of connecting authorship to analytical thinking.

1.2.3   ALEXANDER FLEMMING: OBSERVING THE UNUSUAL

Analytical thinking and writing can only work when scientists and engineers ask open, unbiased questions. When people try to use experiments and data to prove a preexisting opinion, the work is no longer objective. Most scientists and engineers value their reputations for objectivity to be their most important resource.

Alexander Flemming provides an interesting example of a scientist who understood the need for objectivity, because it allowed him to see unexpected results (Brown 2005). Flemming is credited with discovering the antibiotic Penicillin. Prior to 1940, physicians had no way to treat people with microbial infections. Louis Pasteur explained the germ theory of disease in the 1860s, but there was no way to treat humans with bacterial infections. Flemming saw many soldiers die from infections while serving as a physician during World War I.

Antibiotics used prior to the 1940s not only killed microbes, but also human cells. Flemming's initial discovery of lysozyme, an antibacterial compound found in human mucous tissues, convinced him that there were chemicals that could kill microbes without harming human physiology. In the late 1920s he grew cultures of *Staphylococcus* in glass petri dishes, and tested chemicals to search for an antibiotic that would kill the microbe.

In August of 1928, Flemming prepared to go on vacation, and cleared a lab bench for a colleague by simply stacking petri dishes with *Staphylococcus* cultures near the sink. When he returned to the lab in September, Flemming started stacking the petri dishes in a tub of Lysol to kill the bacteria. A few dishes escaped submersion. When a colleague came to the lab, Flemming was busy cleaning dishes, and pointed to the tub. As he picked up a dish, he noticed the *Staphylococcus* colony was contaminated with a mold, and that the bacteria did not grow next to the mold. Suddenly, he recognized the possibility that the mold made a chemical that had antibiotic properties. Had

Flemming not been a keen observer, and open to interpreting unexpected results, he would have cleaned all the petri dishes, oblivious to the evidence in front of him.

The mold Flemming found was *Penicillium*, and he called the antibiotic chemical it produced Penicillin. Flemming published his discovery (Flemming, A. 1929). But it was not until World War II created an overwhelming need for an antibiotic that new technologies in chemistry and engineering allowed for commercial synthesis of Penicillin. Flemming, along with those who developed the technologies necessary to commercially produce Penicillin received the Nobel Prize in 1945.

1.3   THE GOALS AND OBJECTIVES

This handbook will help users become more proficient at analytical writing and thinking. This handbook will be useful for students in courses, for professionals as a reference when classes are over, and for those who choose to study it independently.

The specific objectives of this handbook are the following.

1. **Increase user proficiency in analytical writing and thinking.** Most students and those with professional careers consider themselves logical decision makers, yet find it difficult to convincingly write an argument based on evidence and reasoning. This handbook will help all users produce written products with improved efficiency and skills. Teachers will have a framework for courses they teach. Students and independent users will have a guide for writing term papers and research documents that are suitable for classes and submission for publication. Handbook users in the public and private sectors will enhance skills needed to produce documents and reports.

   This handbook will address issues related to reasoning, the mechanics and format of writing, and syntax, but there will be little emphasis placed on developing grammar skills. Instead, the Appendix in this handbook points readers interested in improving grammar to other resources.

2. **Increase user proficiency in assessing and evaluating information.** As students improve their skills in analytical writing and thinking, they will also improve in the abilities to evaluate other sources of information. Handbook users will more critically evaluate information from journals, popular magazines, newspapers, and the media including radio, television, and the Internet.

   For those engaged in teaching scientific writing, this handbook provides guidelines for objective evaluations of writing assignments. Students will quickly learn that all writing is not of equal quality, even in technical and scientific journals, and especially on the web. Teachers should openly discuss the evaluation of writing in class, so students become more competent judges of information and understand a grading system before they begin assignments. Once students know and understand the grading system, they can also learn by actively evaluating the writing of other students. Evaluating others will help students to judge better their own work.

3. **Facilitate Team Writing.** Professional scientific writing is often done in teams. Examples include multi-authored scientific journal articles, committee reports, and reports in environmental consulting. Learning to function in writing teams is an important skill that includes aspects of social dynamics, leadership, conflict resolution, dividing assignments, recognizing individual contributions, and shared responsibility. This handbook provides ideas for coping with complexities of team writing.

   Even documents with single authors often involve a writing team. For example, an author writing a scientific journal article, a student preparing a term paper, or a graduate student writing a thesis should share a draft with a colleague, friend, or graduate student advisor and committee. By seeking outside readers, the author creates an informal writing team that provides constructive criticism for subsequent drafts. The author typically recognizes those who provide comments on drafts in a section of the article's acknowledgments section.

4. **Control Bias and Plagiarism.** This handbook will help the user recognize the bias that exists in each of us; recognizing our bias is the first step to controlling it. Unchecked bias can defeat attempts at analytical writing, and ultimately erode or destroy an author's credibility.

   Writing is a creative process that builds on ideas from the author, and from others. Authors should give attribution to those from whom you take ideas, evidence, data, and quotations. Analytical writing typically draws from the works of others, and will serve as a source of information and ideas for more writing to follow. This handbook shows users how to use information from a wide range of sources while giving credit to sources. Failing to acknowledge the sources of information used, or plagiarism, is among the most serious offense a writer can commit. This handbook also shows how uncontrolled bias can also result in the construction of unfair tests, and create pressures for mistreating or fabricating data. Failing to control personal bias can result in litigation, termination of employment, and loss of professional stature and credibility.

1.4   THE AUDIENCE

Improving skills in analytical writing and thinking is a lifelong process that leads to professional and personal rewards. This handbook stems from my personal experience that includes more than

20 years as a university professor. Early in my career, I recognized that students learned facts, and how to find facts. However, students needed more experience at using facts to answer questions.

I created the course, *Analysis of Environmental Issues*, and have taught it for more than two decades. Much of this handbook builds on the success of this course. Because I am a biologist by training, many of the examples used in this handbook are from the life sciences. But the general principles outlined are relevant to the natural sciences, social sciences, engineering, business, and those careers that require rational thought for making decisions that commit investments of money and other resources.

Students are excited by the opportunity to become more responsible for their learning in inquiry-based teaching environments and will find this handbook has great utility. Teachers will see firsthand the excitement that comes when students realize the empowerment that comes with the process of analytical writing and thinking. Students change from gatherers of information into those who seek specific information for a purpose. With such a transformation, students can effectively seek answers to life's questions and develop their personal interests in science, and beyond.

This handbook will assist those who teach science courses for students in high school and in colleges and universities, and especially aid teachers of science curricula that require capstone writing courses. Faculty members and teachers in science who are open to improving the writing skills of science students will find this handbook provides the framework for writing projects that span an academic quarter, a semester, an academic year, or longer.

This handbook provides lessons in analytical thinking, but is not a guide to logic, rational thinking, or formal proofs and hypothesis testing. Those readers interested in logic and proof should find further coursework or reading to improve reasoning skills. Although, this handbook presents writing tips, there is no substitute for understanding grammar, syntax, and mechanical writing skills. All professional writers often consult books on grammar, technical writing skills, a thesaurus, a dictionary, and other writing aids. Please see the Appendix for suggested books related to these topics.

CHAPTER 2

Descriptive and Analytical Writing

2.1   DESCRIPTIVE OR ANALYTICAL WRITING?

Analytical writing incorporates information to answer specific questions; achieve goals, test hypotheses, decide new approaches, identify relevant information, and explain the importance of the new ideas. The process of analytical writing uses conventions and formats that help readers, who often must look at many documents and then form ideas for their own writing.

Scientific writing can be descriptive rather than analytical, and descriptive, scientific writing is a vital form of communication. In many cases, descriptive, scientific writing contributes information to magazines, newspapers, and other media forms often accessed by the public. For example, descriptive writing might explain the life cycle of a penguin, list all the species present in a biological survey, or catalogue the sequence of geological formations of rock strata found at the Grand Canyon.

Descriptive writing is useful for explaining what exists at a point in time, and provides entertainment, information, education, and promotes awareness. The stage is often set for descriptive, scientific writing with statements such as the following.

1. "The purpose of this article is to characterize the flora of the Mojave Desert."

2. "This article reviews the literature on space shuttle reentry."

3. "This report will discuss the oil spill damage."

In none of the three examples is there a reference to specific objectives, hypotheses, or questions that will be pursued, nor to significant conclusions. Descriptive, scientific writing can also lead to large topics covered in ways that lack focus, treat each element discussed with equal importance, and appears as a literature review or encyclopedic treatment of information.

Analytical writing typically includes some descriptive writing, especially early in the document. The analytical writer might describe aspects of the penguin life cycle, a biological survey, or Grand Canyon geology to set the stage and rationale for a set of questions, goals, or hypotheses that then become the focus of the writing and thinking. The stage is set for analytical, scientific writing with formative statements such as

1. "Predicting impacts of invasive species on biodiversity of flora in the Mojave Desert."

2. "The report will prioritize the risks of shuttle reentry."

3. "The objectives of this paper are to assess the risks to Chinook salmon egg viability to oil spills in Puget Sound."

In the three examples of formative statements in analytical, scientific writing, the author promises to deliver specific products based on critical thinking about specific information. Analytical, scientific writing typically has specific, focused subjects, consideration of the most important elements and little mention of unimportant ones, and a clearly stated beginning and end-point.

2.2   ANALYZING INFORMATION AND CONCEPTS

Analytical, scientific writing involves actively using information to derive a new understanding, and typically goes far beyond describing sites, processes, or phenomena. The new understanding derives from careful thinking about evidence, understanding limitations of information available, developing assumptions, and identifying just how the new understanding is useful in compelling ways. Analytical writing is necessary to clarify the importance, rigor, and utility of the thinking. Even though all decisions in areas of planning and resource management rely upon analytical writing, some analytical writing may result in conclusions that have no immediate applications.

The process of analysis often involves use of data, but also includes concepts, theories, models, and other abstract forms of reasoning. Such analysis requires writing in just the same way as does a data-centered project.

2.3   THE PRODUCT UTILITY

Analytical writing often provides the foundation for making decisions, and therefore results in intellectual products. In many cases, peer-reviewed journals, using a process of anonymous reviews, publish analytical articles that present advances in recent research. Although the publication of many books and journals requires review, the anonymous peer review is the highest review standard because the reviewer is unknown to the author, and the professional and personal relationships between the reviewer and author are not in jeopardy.

If a reviewer has strong bias about the discipline, or a personal relationship that prevents an objective review, the reviewer must declare a "Conflict of Interest" and not review that specific article. In some cases, reviewers should avoid evaluating all the work from authors where conflict of interest puts the work in unfair light.

The results of analytical, scientific writing published by governmental agencies, are sometimes known as the "Gray Literature." Reports written by staff in local, state, and federal governments are generally available to the public and easily obtained. In some cases, controversial reports may be difficult to acquire, but the Freedom of Information Act allows access to most documents produced with public funding. In general, the pre-publication review of gray literature is not as rigorous as for peer-reviewed journal articles. Gray literature can also be more difficult to find, is less likely to be in libraries, and therefore has less impact in the scientific community. However, gray literature often contains detailed information that exists nowhere else, can be an excellent source of information and ideas, and can often be readily found as agencies put information on their websites.

Scientific, analytical writing produced in the private sector can either be proprietary with restricted distribution, or made available to the public. Corporations and smaller businesses often produce reports and planning documents that may be relevant for internal purposes, but are not of general interest. Consulting companies often specialize in producing reports for a client or customer who lacks the specific expertise to produce reports. Consulting firms that produce reports for a client negotiate in advance to specify the scope of the project, intellectual products, report distribution policies, and other details eventually written in a legal contract before beginning work.

Scientific, analytical writing often results in products, and the development of tools that have specific use. Some products from analytical writing include the following.

1. **Planning Strategy.** Strategic plans provide important tools for making decisions in business, projects, and even personal activities. Organizations, corporations, agencies, and units within them, often write strategic plans to outline their vision for the future, and to clearly state their mission, goals, and functions. Importantly, planning typically involves defining programmatic change for an organization, with inputs, internal processing, outputs, and output assessments.

   Strategic plans can also document the rationale for decision-making based on projected trends and outcomes. The rationale in strategic plans often reflects defined goals, tolerances, acceptable losses, the size of investment, and expected gains of resources, position, or market share. The limits set for resource use in strategic plans require careful thinking expressed in writing so the reasoning used for deciding strategies is clear. Strategic plans also include tactics for achieving strategic goals, and assessment of progress for both tactics and strategies. Analytical writing plays a vital role in all such planning processes.

2. **Defining Criteria.** Establishing criteria defines standards, measures, rules, or tests and provides critical tools for making decisions that reflect analytical thinking. Those in business, science, and engineering often use criteria based on analytical writing and thinking. Organizational leaders may not always agree on the criteria used for hiring, merit raises, employee evaluation, monitoring environmental pollutants, making loans for mortgages, or foreclosing on mortgages. Setting such criteria requires creating, and communicating in writing, an analytical framework that reflects strategic values, collecting information, and drawing appropriate conclusions.

   Setting criteria are common practices, and a good example is student grading. In the syllabus for a high school or university course, the instructor will list the criteria for grading that might include specific point allotments for examinations, writing assignments, attendance, and participation. When students know the criteria for grading, they also know the rules for evaluating student performance.

The process of establishing criteria requires analytical thinking as the instructor connects course objectives to measuring student learning. For example, teachers must determine whether the term paper, or the final examination, has more value for appraising student learning, and award points for each based on the values of learning outcomes established at the beginning of the course. Establishing the grading criteria has a profound effect on the teaching and on student activity during the learning period. Analytical thinking anticipates how the grading criteria will influence the learning experience.

3. **Establishing Priorities.** Setting priorities establishes a rank order of importance for taking measurements or for making decisions about the sequence of activities and actions. Establishing such a rank order for parameters takes considerable analytical thinking because the assumption is that not all activities are equally important. Priorities reflect both values important to businesses and organizations. Ultimately, each of us establishes priorities that govern our lives.

People make personal decisions reflecting priorities when they decide to buy a home rather than a new car. Some homebuyers may give a higher priority to house location over total floor space. Other buyers may see greater value from investing in newly constructed houses rather than older homes that need remodeling. Regardless of which home is purchased, buyers prioritize the qualities of the house, whether they make a formal list, or not. The idea of formalizing the prioritized list, and writing it down, is an important form of analytical writing that crystallizes a rationale for making decisions.

Just as homebuyers establish priorities for a house, businesses and organizations establish and use priorities to guide decisions leading to actions. For example, a business may prioritize the need to increase profits, increase market share, decrease operating costs, or expand into new territory. The decision to place one action above another requires careful analysis of the costs, benefits, and return on investment, with the information carefully packaged in a business plan.

4. **Setting Triggers.** Setting triggers puts pre-made decisions in motion, initiating actions at future set points. Triggers initiate previously considered decisions when specific conditions are met. A common example is a trigger for sale or purchase of stock. A stock buyer might acquire 500 shares of General Motors stock at $50 per share, and set two

triggers. One trigger might be set to sell the stock if it drops to $40, to avoid further loss. Another trigger might be set to sell the stock if it rises to $60, to ensure a profit.

The person who sets triggers and buys the General Motors stock may decide they can afford to lose $1 per share, but no more, and to sell the stock should the price drop more that that. How does an investor decide how low a declining stock must go before selling? A person buying General Motors stock at $50 per share might alternatively set a buy trigger if the stock gets to $40 per share; if the stock was a good buy at $50, it must be a real bargain at $40. Setting the trigger to sell stock at a specific, target price depends on the entire portfolio of stocks, their performance, and losses a person, or investment fund, can tolerate.

Selling stock that is increasing in value can also make for a difficult decision. After all, that is why people buy stock, so that they make a profit when it goes up. But, to make a profit (excluding dividend payments), the stock must be sold. Attempting to sell stocks at their highest value is simply a guess, so another approach to selling is to set a trigger that will define a specific investment objective.

Stock market analysts work to assess stock prices on the basis of corporate values, profits, losses, corporate debt, management team, and projected growth. The assessed value of the stock can then be compared to its actual value, and decisions to buy, hold, or sell can be made relative to an objective assessment. Approaches to assessing values of stock, or any other objective assessment for that matter, can vary widely. So too can the skill and the success of the assessor.

Setting triggers in the stock market, with real estate, with blood cholesterol levels, or any other factors requires analytical thinking that goes beyond guessing or instinct. Naïve investors buying General Motors stock often do so without knowing when to sell it, or when to buy more. Careful planning, analyzing the impacts of a range of performance parameters, and personal values all influence trigger set points thereby minimizing making capricious decisions. When we choose a stockbroker, real estate agent, or physician, we will try to find the person with high prospects for success. Chances are this person will be an analytical thinker and will be able to show you in writing a history of performance, rationale and approaches used, results to expect, and specific recommendations.

5. **Managing Risks.** Those confronting decisions about action plans, resource use, or future outcomes must recognize the existence of risks. When we recognize and acknowledge risks, we can develop approaches to manage them. Analytical thinking and writing

about risk management shows foresight, rigor, and comprehension of uncertainties that affect decisions.

The process of risk management is a discipline onto itself, and embraces a unique set of concepts and vocabulary specific to the field. Risk management often begins with assessment, and ultimately results in strategies for reducing risks to acceptable levels. For example, risk management concepts apply directly to insurance and other industries centered on financial management. Risk management also applies to environmental resource management, and includes setting standards for pollution levels, listing endangered species, and developing resource management plans.

A clear example of risk management is seen in the insurance industry. The insurance buyer spends money to mitigate risk. Conversely the insurance company ensures income from premiums is greater than claims paid.

Each individual who buys insurance is managing risk so that if disaster strikes, he or she incurs an acceptable cost. Specific examples of insurance an individual might buy includes car insurance, household insurance, home owners' flood insurance, medical insurance plans, life insurance, and dental insurance.

Consider common car insurance as an example of an informal risk management issue. Car insurance has options to mitigate risks from collision, liability, health, towing, and roadside assistance. Factors affecting risks to drivers include age, gender, the kind and number of cars involved, and the amount of money available for car insurance. Analysis shows single males under the age of 25 pose greater driving risks than older or married males. More specifically, insurance companies manage risks to their own resources by charging single males greater premiums in anticipation of their greater claims. Similarly, a driver of a fast expensive sports car will pay greater premiums in anticipation of claims greater than those for the driver of a mid-sized sedan. So, insurance companies are managing the risks to their financial wellbeing by adjusting premiums to claims.

Even though those in households do not typically write formal reports evaluating risks before buying insurance, people often gather information to compare insurance costs and coverage in an effort to manage risks. On the other hand, businesses and organizations also purchase insurance using analytical writing as the underpinning to justify a complex investment decision involving existing and planned resources, corporate values and responsibilities, and estimating risks.

6. **Making Predictions.** Predictions forecast the future and the tools for doing so require rigorous analytical writing and thinking. Common predictions include weather fore-

casts, rates of inflation and unemployment, and prognoses related to health issues such as cancer, diabetes, or high blood pressure. Scientists and engineers use predictions to estimate melting rates of the polar ice cap, rates of extinctions, genetic drift, and the outcome of management plans for ecological systems.

Tools also exist for estimating future conditions. In some cases predictive tools use historical databases to extrapolate based on a trajectory for past changes. Predictions can include interpolation by making a forecast between two or more existing data points. In other cases, predictive tools use theories to create models that predict future conditions. In either case, predictions require a thorough discussion of assumptions, the strengths and weaknesses of the modeling approach and mechanics, estimates of accuracy and precision, verification of predictions by independent methods, and measures necessary to improve predictive capacity.

Predictions always carry a measure of uncertainty. There are tools for measuring uncertainty, which can become as important as the prediction itself. A good example is the prediction of how increasing greenhouse gas concentrations impact climate change. Because reducing greenhouse gas emissions, and adapting to climate change, requires huge investments and changes in lifestyle, the issue of uncertainty in the prediction of greenhouse gas impacts provides opportunity for debate. On the one hand, if there is great certainty that greenhouse gas emissions will accelerate climate change, many will want to reduce emissions. On the other hand, if there is a high degree of uncertainty that greenhouse gas emissions affect climate, then there is a less compelling case for efforts to reduce emissions.

CHAPTER 3

Guidelines for Students and Teachers

Teachers in K-12, colleges, and universities are typically comfortable presenting material for students to assimilate. The comfort comes from control of the material used in the class that is often connected to the teacher's academic background. Most of us in education learned our craft in classes taught with teachers providing information in lectures and laboratory work that is carefully packaged to fit within lesson plans and academic time units. Evaluating student progress and learning in classes is commonly accomplished with graded examinations and, in some instances, term papers that are submitted at the end of the course.

Teachers are typically comfortable giving examinations, but term papers often pose difficulties. Science teachers are not typically taught how to structure and grade term papers, so the products from students are often uneven in quality of thought, poorly edited, difficult to read, seem to wander, lack adequate referencing, and evaluation is time consuming. No wonder that science teachers are not fond of assigning or grading term papers.

The following sections cover a few guidelines for teachers using this handbook. Chapter 13 provides examples of classroom activities suitable for both high school and college students.

3.1 INQUIRY-BASED LEARNING

Let the student decide on the topic for writing. Since students must obtain the subject matter for the writing effort, let them choose the topic. A process that works is to let students choose a general topic with the understanding that they will submit a title to you. The title will give the instructor a chance to be sure the writing is appropriate in scope and will begin the process of critically thinking through the organization of the paper. Students working on self-selected topics will put large efforts into producing analytical, scientific writing because of their personal interest in the topic and the responsibility that comes with ownership of the project.

3.2 SUBJECT AUTHORITY NOT NECESSARY

Teachers will not be, cannot be, and should not be an authority on every topic students might select. In general, science teachers have backgrounds broad enough to appreciate and be conversant in science topics that are peripheral to their training. Teaching analytical, scientific writing is not so much about teachers explaining difficult science facts to students, but teaching students to write for an educated audience. The teacher's goal should be to encourage students to become authoritative on their topic, and to write a paper that is interesting to a person with a scientific background.

Teachers in the sciences, engineering, and business should develop skills and approaches to teach writing, but need not be authorities on all subjects covered by the writing from students.

3.3   PROVIDE GUIDANCE NOT ANSWERS

Students will invariably seek help from the course instructor. Requests for help will begin with the selection of titles. Students will also ask where to find information for the paper, how to contact experts in the field, and how to analyze data. Teachers will be tempted to give specific answers: This is the best title, go to the library for information, call the U.S. Environmental Protection Agency to speak with an expert, and use linear regression to predict. In general, teachers and instructors should provide general guidance and present options for students. However, teachers and instructors should typically avoid giving students specific answers to questions or solutions to problems. Providing too much support can result in teachers taking control of the project from students, and worse still, makes you part owner of the project. Instead of providing answers, provide guidance or a framework that students can use to answer their own questions. Make sure students know that there is no "correct" answer to their question, that they have an array of options, and that they are responsible for the choices.

3.4   SOME STUDENTS WILL STRUGGLE

Students, like instructors, are most familiar with a lecture format for classes. Choosing topics for extensive work, and creating a piece of analytical writing will challenge students. As with any learning situation, some students will grasp the challenge more quickly than will others. The process of analytical, scientific writing will distribute the range of student performance more broadly than many other kinds of learning and teaching activities. Students who quickly come to understand the goals of scientific writing will be obvious, and will excel. Students who struggle will become obvious. Instructors working with students who need help will see those who initially need help will suddenly grasp the challenge, take off, and work rapidly and productively. All students will improve and show increased ability to work independently.

3.5   CHECKPOINTS ARE NECESSARY

Teachers cannot assume that students who outwardly appear to be doing well are really excelling. Nor can teachers grasp the issues that all students will be facing in the early stages of writing, and the other phases that follow. Build instruction so that students must regularly submit products leading to finished papers.

Checkpoints can be classroom activities. Students can individually make reports to the class in short presentations. Early in the course, each student in turn can explain to the class the title of their project and their objectives. Later in the course, students might explain the approaches used to accomplish objectives. Such presentations in class not only provide checkpoints for students, but also give them experience in communicating and explaining their progress and their problems.

3.6   LET STUDENTS SHARE AND EVALUATE WORK FROM OTHER STUDENTS

Students are wonderful teachers of other students. Provide opportunities for students to share and evaluate the work of other students. Students with chances to talk with others who are coping with the same issues, but from different topics and perspectives, will often come to solutions that could never otherwise be developed. Surprisingly, many students will be more critical of work from other students than are instructors.

3.7   BECOME COMFORTABLE WITH UNCERTAINTY

Checkpoints for progress will provide evidence of student progress. However, there will always be elements of uncertainty. Students may not know the results of analysis until late in the project, even if the analysis involves simple graphs, tables, or statistical methods like regression, correlation, or means separation techniques. Students may also have experiments to evaluate, polls and surveys to take, GIS analysis, or models to validate and run. Answers will be unknown, and that is exciting for everyone. Much of the excitement of discovery is the anticipation, and teachers and students should learn to enjoy the uncertainty, speculate on outcomes, and consider the implications of results.

CHAPTER 4

Choosing Topics

4.1 WRITER'S CHOICE

Writers write best when they choose their own topics. When writers feel they have the freedom to choose a topic, they have ownership and an obvious, built-in interest. Writer ownership and interest are necessary because of the extensive effort necessary to produce analytical writing. When teachers give topic ownership to students, the action can generate the momentum student writers need to read, think carefully and outline, write, and rewrite.

Instructors may think allowing students to choose topics will result in students owning the course. However, even when students choose their topics, the instructor still owns and controls the pedagogical process of teaching analytical writing. More specifically, the instructor still presents the format for learning, leads interaction with students, advises on student activities during the class, and evaluates student products.

Instructors typically have no way of knowing students well enough to make good writing assignments for them. Almost all students and writers have unique interests, backgrounds, or access to important information. Building on the unique qualities that writers bring with them will lead to new ideas and insights, not only for the instructor but also for the writer. However, in some cases, teachers and professors might identify a large area within which students must choose their topics. Examples of large areas for class focus include energy, environmental issues, nutrition, agriculture, accounting, etc.

Instructors may also worry that students will choose topics for which neither the instructor nor the student have expertise. Even so, allowing students to develop such topics can result in an important educational experience for all involved. Even when instructors do not know an area of science, they can teach the process of analytical writing, and develop the student's ability to write to an audience of an educated public.

In the end, students should be able to explain their project to an instructor, professor, other students, and to those with a basic level of scientific expertise. Analytical writing should be at a level appropriate for a wide range of readers, including those with diverse scientific backgrounds.

4.2   THE ANALYTICAL SET-UP

Choosing an analytical topic is a complex matter, even when the writer knows about the general area to be explored. There are two major pitfalls:

1. **Topic Breadth.** Writers often choose a topic that is too broad, with the outcome that the writing lacks focus and development. Writing topics that are too broad also lend themselves to descriptive writing. For example, a student writing a paper entitled, "Oil Spills: Ecological Impacts" will be writing an encyclopedic paper describing all the ecological impacts of oil spills that have occurred, and all the possible ecological impacts. The paper will have no logical beginning, context, or end point. Books are written on the ecological impacts of oil spills, and the topic will certainly be beyond the scope of most students and other writers.

   Topics that are too broad in scope for students can limit progress. For example, most authors cannot write all there is to know about oil spills, but they can write about specific aspects of ecological impacts of an oil spill event. Alternatively, authors might be interested in oil spills and sea birds, or dispersion of oil spills. The point is that students and other writers can unwittingly set themselves up for descriptive, long papers that are winding literature reviews with no conclusion. If so, students may miss the opportunity to develop focused writing that has purpose, is direct, and answers specific questions. Examples of more focused titles for papers on the ecological impacts of oil spills include:

   ○ "The Exxon Valdez Oil Spill: Comparing Vulnerability of Sea Birds and Shore Birds."

   ○ "Predicting the Next Oil Spill in Prince William Sound: Analysis of Shipping Traffic, Weather, and Currents."

   ○ "Prioritizing Risks of Oil Spill in Prince William Sound during Winter Seasons."

   Writers often feel that they must select a broad topic to ensure that there will be enough information to write a paper of adequate length. Most young writers do not understand how quickly a paper that initially appears narrow in scope can grow into a substantial project. Checking the titles of theses in university libraries, of articles in peer-reviewed journals, or in the gray literature will quickly reveal that writing with sharp focus and analytical thinking results in substantial manuscripts.

2. **Descriptive Topics.** Writers often seek the comfort of descriptive topics. Examples of descriptive writing include surveys, literature reviews, and are often characterized by titles that lack focus. For example, a literature review summarizing each article written

on a topic seems a sure bet to a writer; if the topic is big enough, there will be plenty of literature to describe. Literature reviews often lack critical or analytical thinking, and can be recognized because each piece of discussed literature is considered equally important. A descriptive literature review might offer some critical thought about each piece, but falls short of synthesizing the importance, direction, or implications of a large body of work involving many writers over a prolonged period of time. Rarely are descriptive literature reviews capable of including all literature on a topic, even with the author's best attempts.

A descriptive article on the "Eastern Deciduous Forest" might include lengthy explanations of all the forest biomes, and a substantial page or two on each dominant species. Although a descriptive paper on the "Eastern Deciduous Forest" might be relatively easy to write because there is so much information written about it, such descriptions are not likely to be useful. Foresters cannot use such a document to help manage forest resources, and such writing is not of much use to those wanting to learn about recreation potentials, the most important medicinal plants, or other aspects of forest ecology. In short, there is not much market or utility to descriptive writing.

4.3   CHOOSING FOCUSED, ANALYTICAL TOPICS

Writers can initially choose broad topics connected to an interest. In the early stages of writing, the topic will be open ended, without conclusion, and lead to further reading, thinking, and discussions with the teacher, experts in the field, and other students. With reading and review of subject matter, students will begin to see questions, issues, and opportunities for their project. The process of reading will move students toward the more interesting material, largely because innate curiosity leads to interesting questions, emerging issues, and the space where questions exist, but answers are not yet known.

The initial topic of "Oil Spills: Ecological Effects" starts with a large stack of books and articles, but quickly migrates because of the reader's interests, background, and ambition. Prospective authors read in their subject area with in an eye for detail and questioning that is quite different from other types of reading.

The author who reads in a subject area prior to writing will, for example gather information on oil spills and ecological impacts. If the author is interested in biology, the literature on this topic will accumulate, and the student's interest may move toward reading about a specific oil spill and efforts to help soiled birds. How do ornithologists rate the various approaches to cleaning birds? Which species respond best to cleaning? Are some birds more sensitive to oil, and what physiological attributes confer resistance to contact with oil? The answers are in the printed material, and bits and pieces of information can be found in a variety of articles. As the author sharpens the topic, the initial title of "Oil Spills: Ecological Impacts" is taking the direction necessary to become analytical writing that has strong purpose.

The reader who is author of a paper on "Oil Spills: Ecological Impacts" may choose to focus on oceanography. If so, the initial phase of reading may lead to collecting and reading articles and papers on the role of currents that cause oil spills to spread and how ocean temperature and wave action affect volatilization and evaporation of oil on water. Are currents or evaporation more important for dispersing oil spills? How important is temperature for evaporation on a flat surface? How about evaporation with wave action? Can models of current, water temperature, and wave action be used to calculate the fate and dispersion of oil spills?

The process of choosing a topic is dynamic because the topic evolves in the writer's mind. Typically, the topic moves from a large, general area of interest becoming more refined and focused as the author gains more insight from background reading. During the first week or two of reading, the topic should quickly move from a general area of interest to a well-contained set of themes, and become more tractable as the author considers the time and resource material that is available. In non-academic settings, the topic should also mature as the author reads, thinks, and develops ideas.

4.4   ORDERING PIZZA: AN EXAMPLE OF TOPIC DEVELOPMENT

The process of ordering a pizza can help show how analytical thinking connects to a common life event. The example of ordering pizza, more importantly, shows how analytical thinking develops to expand awareness. As the thinking behind the ordering unfolds, the change in scope of thinking can also be connected to writing. Of course, people would never write about the progressive thinking involved in ordering pizza, or would they?

Imagine you and three friends go to Dominico's Pizza Parlor to order pizza for dinner. You all agree you want the "Supreme Combo" and the question simply is which size to order. Prices* for the "Supreme Combo" are:

- Small, 10-inch diameter: $10.00

- Medium, 12-inch diameter: $13.00

- Large, 14-inch diameter: $14.00

- Extra Large, 16-inch diameter: $17.00

  *The sizes and prices are taken from an actual pizza menu. In scientific writing we should always use the metric system. However, the demonstration uses the standard units of inches.

The question for the four friends is simply to decide which size pizza to order. One way to decide is to look quickly at each other, have a quick discussion, and let one in the party tell the waiter, "We'll have a Large, Super Combo pizza." In this case, not much thought is in the decision.

Alternatively, someone in the party of four might say, "Gee, how much pizza will there be if we order a Large, Super Combo?" One approach would be to quickly calculate the circumference of the pizza with the equation:

$C = \pi D$

Where

C = circumference

π = pi = 3.14

D = diameter

Solving the equation will allow the four friends to find the total circumference of the pizza, divide the circumference by the number of people, and see if that seems enough food. So, the circumference of the Large Super Combo pizza is:

C = 3.14 x 12 inches in diameter

C = the whole pizza measures 37.4 inches, around the outside crust

With four diners, that would be roughly:

To determine an equal share for each diner, find

37.4 inches/4 = 9.4 inches per person

If the diner chose to eat their share of the pizza in three portions, each person can envision getting three slices of pizza, each about 3.1 inches around the outside crust, and about 6 inches (half the diameter = radius) long on each side.

Now the question becomes, is the Large Super Combo pizza too much, or is it too little? Consider the diners. If the four diners are students trying to watch their weight, the pizza might be too large. If the diners are four athletes trying to gain weight, the pizza may be too small. In either case, the make-up of the dining group is important, but with a brief step in analytical thinking, people can start to get a pretty good idea of whether the Large Super Combo pizza is the right size.

What if the four diners are college students with limited income, and they want the most pizza for the money? The students know they can take left over pizza home for later meals. The motive is getting the best deal, or the most pizza per dollar. The students know to determine the area of a pizza:

$A = \pi r^2$

Where:

A = Area

π = pi = 3.14

r = radius

Solving the area for each pizza size and dividing by cost will give the cost per square inch of pizza, which is a great way to find the most pizza for the money. The result is:

Small: A = 3.14 $(5)^2$ = 79 square inches

$10/79 square inches = $0.127/square inch

Medium: A = 3.14 $(6)^2$ = 113 square inches

$13.00/113 square inches = $0.115/square inch

Large: A = 3.14 $(7)^2$ = 154 square inches

$14.00/154 square inches = $0.091/square inch

Extra Large: A = 3.14 $(8)^2$ = 201 square inches

$17.00/201 square inches = $0.085/square inch

The simple arithmetic shows that the cost per square inch of pizza ranges from almost $0.13 to less than $0.09. Put another way, the Extra Large Super Combo pizza is nearly 50% cheaper than the Small Super Combo pizza, on an area basis. The decision on which size gives more pizza for the money is now an easy question to answer.

The pizza story to this point explores the issues of size and cost. What other issues could be important? Of course, the issue of quality should come up. In this example, the diners chose to go to Dominico's Pizza Parlor. Important factors in the choice of pizza parlors might be not only the quality of the pizza, but also the location and atmosphere of the restaurant.

In consideration of pizza quality, the four friends likely tried many pizza parlors, and generally agreed that the Super Combo at Dominico's was the best available Super Combo type pizza. Or, perhaps the four friends evaluated pizzas from different restaurants by using a rating system for common criteria, such as crust, sauce, and toppings. On a 10-point scale, with 10 being perfect, the results may be:

Dominico's Super Combo Pizza

Crust = 9

Sauce = 8

Toppings = 9

Average = 8.7

Alphonso's Super Combo Pizza

    Crust = 7

    Sauce = 7

    Toppings = 8

    Average Score = 7.3

Ramona's Super Combo Pizza

    Crust = 7

    Sauce = 6

    Toppings = 8

    Average Score = 7.0

After such a survey of pizza restaurants, the four friends concluded that the quality of the Super Combo pizza, based on analysis of selected criteria, was the best choice. Of course, the analysis could become more complex. The friends could add criteria to evaluate not only the pizza, but also the convenience of the location, the atmosphere, the beer selection, and the average waiting time and quality of service. In addition, weighting criteria places greater importance for some criteria, such as toppings and lesser importance to other criteria such as quality of service. A more complex point system could add refinement to statistical analysis. A resampling schedule might be important to check for consistency of the pizza quality or to assess changes in quality that might come from change in ownership, chefs, and wait staff.

Is the pizza story over-developed? Some might think, "Wow people going out for pizza will never consider all the issues of quantity per person, price per unit area, and objectively comparing restaurants." On the other hand, owners of pizza restaurants go through just the kind of thinking and the calculations from above to create the best return on their investment dollar.

Analytical thinking guides decisions about how much capital to invest in a location for a pizza restaurant (or any business), to accurately approximate costs of staff and cooking needs, to cover direct and indirect costs, to forecast the revenue flows as the restaurant begins operation, to project what operating benchmarks are necessary for the business to succeed, and the rate of return on the capital needed to justify investment. A business plan must show how many pizzas will be sold and their price, and how the income from sales (of pizza and all the other offerings), will generate the revenue to pay back the investors, pay operating costs, and provide jobs and incomes for the owners and workers.

Restaurant owners, and those investing with them, will do a careful business plan before committing resources to a venture. If either those who want to find investors, or the potential investors, do not have the expertise or time to make the plan, they will likely hire consultants who will do the work. In any case, before those who start a pizzeria can persuade others to invest in the idea, someone will write a document that addresses the costs, benefits, risks, and investment objectives connected to a business proposal and plan.

If restaurant owners go through analytical thought processes before starting a business, so too do the customers. The degree to which potential customers analyze their pizza orders will of course differ with each experience. If four friends are going for dinner, there may not be much thought. However, if you are ordering pizza for a large party, you might think more carefully about the amount of food to purchase, price, delivery, quality, and other factors.

Thinking analytically about the pizza order does more than provide an example of how a topic can develop. The example also demonstrates that analytical thinking, and communicating the thinking in writing, can provide a skill that has market value and can result in a job. There will be a need for those who can write the business plan. There is also an opportunity for restaurant critics to write objective, analytical reviews of pizza parlors, and other restaurants.

CHAPTER 5

Writing Teams

5.1 A COMMON PRACTICE

Learning to write in writing teams is important for many careers, and requires skills beyond those of single-authored documents. Analytical writing is often the effort of several writers that form a writing team, and examples of analytical writing produced by several authors are more common that are single-authored articles. In fact, most scientific journal articles have several authors, with some articles having long lists of authors. Environmental consultants, engineers, investment managers, and many others in both the private and public sectors often produce reports by writing in teams.

Rarely, if ever, does analytical writing emerge from one person writing as an island, alone, and without the involvement of others. For example, even when a single writer creates analytical writing, the author must build on the writing and ideas of others who receive citation or attribution. Admitting the role of others does not take away from recognition that comes from writing. On the contrary, authors who relate their writing to ideas of others will usually increase the interest level from readers, and expand the number of readers. More simply put, attributing ideas from others with citations and acknowledgements is the right thing to do, and allows the author to join a community or team that is making progress in a specific area of thought. Single authors may actively seek input from others, provide attribution, and in so doing create an informal writing group.

Formally established writing teams often produce analytical writing. A working group manager or instructor may form writing teams by simply making assignments. Alternatively, writing teams can be formed when units in government or business ask for volunteers. Whether by assignment or volunteering, team writing generally follows one of two schemes.

1. **Piece Assignment.** Each member of a writing team assumes authorship responsibilities for a specific chapter, series of chapters, or specific sections of a document. The assignment includes creating the first draft, making revisions, and delivering a final team report with elements of writing integrated into a single document. Regardless of the piece assignments, the process of integrating elements to make a final document is an essential final step.

2. **Role Assignment.** Each member of the writing team assumes responsibility for a specific task needed to complete a document. One or more persons may be responsible for creating the first draft, others may manage the revisions and editing, and still others may

produce the tables and figures. The process of integrating the activities of those with specific roles is critical to completing a cohesive document.

One advantage of writing teams is they can produce work more quickly than single authors. Large, analytical documents often have completion deadlines and writing teams collaborate to create timelines that single authors could never meet. For example, an engineering firm may have four months to prepare a comprehensive proposal for building a water treatment plant for a city. The proposal will have extensive narrative including site description, technical designs, environmental impacts, project management, budget and cost justification, and application forms. A writing team will prepare the proposal by dividing the job into manageable units, creating a first draft, making revisions in response to comments from outside readers, and creating a final, polished proposal that might be several hundred pages in length.

Interdisciplinary writing teams can also bring a wide array of expertise to analytical writing. An environmental consulting company might have a contract to produce a report on the impacts of a gasoline tanker that spilled fuel into a stream. The report might include analytical writing from plant ecologists who assess the riparian vegetation. In addition, phycologists may survey populations of algae. Microbiologists, entomologists, fisheries biologists, and aquatic chemists might also contribute to analytical writing in the single report.

5.2   CONTRIBUTING TO A WRITING TEAM

Writing teams are complex, social groups that can bring all the joys and sorrows of shared endeavors. Writers are not born with all the skills necessary to be strong contributors to a writing team. Becoming a valuable writing team member is typically a learning experience that goes beyond the basic skills of simply writing as a sole author. Ten guidelines for being a good citizen on a writing team include the following.

1. **Accept assignments or roles.** If an instructor provides an opportunity to participate in a team writing assignment, accept the chance to develop skills that will be important once you leave the academic life. Accepting such assignments in the academic setting is perfect preparation for the moment when an employer or supervisor assigns you a team writing assignment. Welcome the opportunity to participate in an exciting effort to advance thinking, help make decisions, and contribute to the network of thinking in your unit.

2. **Be credible and follow through.** Your credibility as a student or employee is your most important asset. Protect your credibility by making commitments, and fulfilling them. Once you accept responsibility for a part of a team-writing project, you must follow through. Authors should be aware of deadlines for specific stages of work, and make

the deadlines relevant to assigned responsibilities. Attend all scheduled meetings, and when attendance is not possible, let others know of your pending absence, in advance. Authors who know they will miss a meeting must make arrangements so they are present in concept and will learn meeting outcomes. If authors cannot make a deadline for a commitment, let others know, and when possible, adjust the timeline. Such adjustments are best made far in advance of a deadline. Requesting a change in delivery date well before a deadline is much better than missing a deadline.

3. **Find ways to make constructive, positive suggestions to the team.** Good writing team participants make positive contributions during planning meetings. Participating authors generally have the background and skills to complete their assignments. Make positive suggestions so they reflect enthusiasm for the effort as a whole, for each role in the team, and for the expected products that will come from the writing. Contribute ideas to the writing team that shape the writing. Feel free to discuss the contributions from you, others, the deadlines for products, and the vision for the final document.

4. **Find ways to listen, acknowledge, and accept ideas from others on the team.** Two heads are better than one, and there is usually value in collective knowledge and wisdom. Listen carefully to ideas from others. Whether in agreement or disagreement, other perspectives, disciplinary backgrounds, and values should influence each authors' thoughts and opinions. Listen most carefully when in disagreement with an issue or perspective. Make an effort to understand reasoning that runs counter to other opinions. Acknowledge others who present ideas, and make sure they understand values from other authors on the team. Do not be reluctant to embrace the strong, constructive ideas from others. Showing respect for new ideas is necessary for strong team players and facilitates teamwork.

5. **Point out potential pitfalls for yourself and others.** Authors may be in the process of developing a team-writing project when one person sees a pitfall not apparent to others. For example, a once reliable data source may suddenly become unreliable. Personal issues can emerge affecting the quality or size of an authors' contribution, or an analysis as part of the project may produce surprising results that make some of the planned effort unnecessary or irrelevant. Those who see such pitfalls should communicate the issue to the writing team so that everyone can understand the situation and the team can discuss options.

6. **Share sources of information with others.** As writing team members gather information for analysis, an author may come across a source of information that is useful to others on the team. Be sure to share relevant, important information sources with others;

authors who withhold information and keep it for themselves are not good team members. Withholding relevant information from others on the writing team is detrimental to the whole project. Writers on teams should share information sources that include new concepts or theories, data sets, one or more articles, or even interviews with experts. Sharing information enhances understanding of the information, leads to greater synthesis in the final product, and contributes to team chemistry.

7. **Understand the assignments and responsibilities of team members.** Be clear about the responsibilities and assignments for each writing team member. Start with each team member knowing and owning assignments. Each author should know the plan for producing team products, and for outlining specific topics, chapters, sections, and all phases of the writing process. Team members must know the schedules and locations for writing team meetings, and the deadlines for each writing product.

   Team members must know the responsibilities of the other writers so that everyone can understand the project goals and its development. Pay particular attention to the work of those with topics and responsibilities that lie close to each other. Take note if some members of the writing team are excelling, and others are struggling. Try to find the reasons for the successes and failures of others to enhance team progress. Be constructive when pointing out any critical areas of responsibility forgotten or ignored by others. If appropriate, offer to help others on the writing team if they are stuck and help them work through barriers; conversely, ask if you need help.

8. **Develop a vision for the final product.** All team members must envision the appearance of the final document. Team writing products might include a term paper for a course, a manuscript for submission to a journal, a report produced for distribution to a committee, a proposal for submission to a bank or a funding agency, or a hard cover book. Each of these products requires a different format and level of effort on the part of the writing team.

   One approach to developing a vision for the final product is to have a model document. In the examples above, a writing team might want to see other term papers, articles accepted by a journal, a committee report, a funded grant proposal, or a relevant book. A model provides confidence to the writing team members that the effort will build a substantial product, and allow each person to have a sense of perspective for their writing assignment.

9. **Provide and expect constructive criticism.** No written document is ever perfect, and documents only improve with rewriting and editing. Rewriting and editing should con-

tinue until the writing project runs out of time or other resources. Analytical writing may also stop when the project reaches the point where large efforts bring relatively small improvements, and the project reaches the point of diminishing returns.

When creating a piece of writing for a team project, authors should get feedback on their writing by sharing manuscripts with others on the team. Expect comments on grammar and syntax, but be especially grateful for input on the quality of organization, content, and the graphic material. Authors should expect that others will provide both large and small suggestions to improve writing. Expect critical readers to provide ideas to make the writing more rigorous, relevant, and compelling. Receiving constructive criticism from someone on the writing team will be important for the development of the project as a whole.

Once the team assembles the entire document, expect that it will also go out for review and comment. Some readers may be close to the group that produced the writing, and there may also be outside readers. In some cases, reviewers may be unknown and anonymous to the writers. The quality of reviews from anonymous reviewers is considered most stringent, and most valuable.

Writers should always consider comments from reviewers and readers carefully, but never surrender the role as author. Authors may not agree with some of the suggestions provided, and some of the suggestions may not be good ones. Reviewers may differ in their opinions. For example, one reviewer may ask for more focus, when another will ask for more breadth. The best practice is to be grateful for all reviews, balance the merits of the suggestions, pursue the good ideas, and leave the other ideas to rest. Trying to accommodate all the suggestions from reviewers can be unnecessary, taxing, and give ownership of the work to reviewers.

10. **Expect to receive credit for work.** A common fear when forming a writing team is that each author anticipates doing more than necessary to fulfill their assigned responsibility. Even worse, authors on teams may suspect they will not receive credit for the work done. Most team writing projects will stipulate the names of the authors on the title page, as is often the case for science journal articles and reports. So, the expectation is to identify each writer on the title page as an author.

However, there are some examples of team writing projects without authors. If so, there is generally good reason and a mechanism that allows employers to acknowledge those who make substantial writing contributions. In some cases, an article may involve so many writers that the list of names will not fit on the title page. When the author list is long, or the organizer of the writing team and editor of the article becomes the author, acknowledgement goes to the writers. In addition, some reports in the public and private sectors, research and business proposals, and white papers may not show the names of some or all of the authors. When authors do not appear on the title page, the author list should appear elsewhere in the document.

The balance of writing assignments among team members is never perfect. Some on a writing team do more than expected, and some do less. Those team members who do more than expected may receive rewards such as a better course grade, or a bonus at work. More importantly, part of being a good team citizen can involve putting the need for recognition aside in the short term, for the benefit of the team. In so doing, writers will build their abilities, achieve recognition from those on the team, and find personal reward.

CHAPTER 6

Organization

The organization of analytical writing is critical to developing a well reasoned, concise, compelling document. In addition, the writing should have organizational structure so the reader can find specific elements of the narrative that show the rationale and objectives, reasoning used to answer questions, answers to proposed questions and issues in a clear way, and the significance of results. Authors can write analytical documents in one of several common conventions understood by both authors and readers. The use of a common convention provides documents with predictable structures allowing readers to quickly find specific information within the article.

The organizational structure of a thesis is similar for undergraduate and graduate degrees, but may differ in minor ways between academic institutions. Those who are writing theses should become familiar with institutional requirements long before beginning the writing process. Theses may require unique features such as a signature page for the thesis advisor and committee members, a complete list of tables, a complete list of figures, and specific requirements for appendices. Thesis guidelines will also stipulate format requirements including font, margins, citations, references, and graphic material.

Solicitations for grant proposals often specify deadlines and detailed requirements for proposal structure and organization. In many cases, the proposal solicitation will define the organization and structure of the proposal, specifying the written sections, and content within sections. Requiring specific guidelines ensures that all those submitting proposals provide information relevant to the funding agency, and that proposals have comparable features including content, development and length, expertise, and budget. Funding agencies may choose not to review or fund proposals that do not comply, so authors should study the proposal announcement or solicitation as soon as possible to become familiar with writing requirements and deadlines.

Scientific papers, term papers, book chapters, and articles submitted to journals often share general conventions for organization and structure that are more flexible than for a thesis or grant proposal. Although the organization of papers, chapters, and journal articles is somewhat flexible, using a general convention greatly facilitates both writing and reading. Even so, students should check with instructors to determine organizational requirements. Those editing books with chapters from multiple authors may also stipulate organizational requirements to provide continuity and synthesis within the volume. Finally, journals also standardize organizational and structural elements, and publish Instructions for Authors.

Whether writing a thesis, proposal, term paper, or journal article, a common organizational structure includes: front matter, introduction, approach and methods, results, discussion, acknowledgments, and appendices. Writers familiar with this organizational structure will be able to write in any formats.

6.1   FRONT MATTER

The front matter is the first point of contact between the readers and the author, and since first impressions are important, authors should invest time developing this part of the manuscript. Unfortunately, overlooking the front matter of a paper or analytical writing project is too common among inexperienced writers.

Analytical writing typically has a sharp, focused title. Writers play with titles and change them often through the writing process. Changing and editing titles during writing shows the development of thought and understanding, and that writing is evolving. As the writing matures, so does the title.

The title page not only includes the title, but other information important to the reader. The title page should include the submission date for the article, the author, or author list, and their contact information. The title page for proprietary information, such as a thesis or a book, should include a copyright symbol (©) ensuring that legal ownership of the writing belongs to the author. Copyright is not typically an issue for term papers, whereas authors publishing journal articles or books assign their copyright to the publisher.

Analytical writing may include a table of contents to help the reader find specific sections or information. Some forms of analytical writing, such as a thesis, require a table of contents, whereas other forms of writing may not. A table of contents should identify the main sections of the writing and the page numbers. The table of contents may show the page numbers for major sections and subsections of the writing that correspond to headings and subheadings in the text. Too much detail in the table of contents can make it difficult to use.

Books may also include a preface written by the author that introduces the general idea behind the book, or the events leading to the book, and may also include acknowledgements. A foreword is a similar piece of writing, written by someone other than the author.

An abstract, or executive summary, in the front matter is the part of the article read most often and carefully. The content of the abstract is important because it often determines whether the reader will continue reading, or will make a note to review it in more detail at a later time. The abstract should generally fit comfortably on a single page, be written in two or three paragraphs, and leave the reader with a complete understanding of the content of the document as a whole. The abstract should explain the rationale and approach for the project, summarize the results, and discuss the importance of the work.

6.2   INTRODUCTION

The introduction is a critical piece of writing that sets the stage for the entire article. The introduction must instantly draw the reader into the writing. The introduction need not be a long part of a manuscript, but it should have sections that connect to each other. A well-crafted introduction should include at least three components that 1) set the stage and playing field for the reader, 2) provide compelling rationale, and 3) identify specific goals and objectives. Subheadings should subdivide the introduction into its parts.

The writing in the introduction should build ideas, with broad issues leading to focus, rationale, and the logical development of specific goals and objectives. By the time readers finish the introduction, they will know the importance of the topic, the scope of ideas considered, and the ideas and other products that emerge with further reading.

6.2.1   SET THE STAGE

The writing in the introduction should quickly capture the reader's interest by clearly explaining the subject of the article, and why it is important to the reader. The introduction should highlight the importance of the topic, and convince the reader of the value reading the article. Authors should ask, "Why should a person spend the time to read this article?" If the first paragraphs do not easily answer such a question, consider re-writing to better set the stage. The first section of the introduction can explain the final outcome of the article without detracting from the importance of investing more time reading other parts of the article.

Summarize the vast amount of work already done in the field, but avoid writing a literature review. Literature reviews typically attempt to describe the content of all the articles written in a field of study. Literature reviews are difficult to include within an article because they often are encyclopedic without critically assessing each article. Instead of writing a section that is a literature review, try to capture the most significant and exciting information.

6.2.2   DEFINE THE PLAYING FIELD

The introduction must set the the boundaries and lead to a compelling rationale for the article. Consider writing an article about oil spills. Certainly the topic of oil spills is far too broad for a focused, analytical paper. To narrow the scope to a manageable size, the author should focus on a specific part of the oil spill issue, such as biological impacts of oil spills, the causes of a specific oil spill, predicting the next oil spill, or some other aspect of oil spills that is meaty enough for development.

Authors should explain the reasons for narrowing the scope, and why the project is still a fundamentally important topic. If the article highlights the impacts of an oil spill on sea birds, explain why the choice of birds as the focal point. Perhaps sea birds are the organisms most sensitive to oil, perhaps the Endangered Species Act lists sea bird species, perhaps oiled sea birds attract more research funds than other organisms, or perhaps sea birds are charismatic and attract large amounts of public interest. Regardless of the reason, explain the rationale for the focus. If appropriate, explain how the focus of the writing provides for a basic, important understanding and far-reaching insight. The analysis might represent a model of an approach for use elsewhere, or have theoretical and predictive values that broaden the scope and number of readers while still retaining critical focus.

A central challenge for the introduction is focusing background information in a large subject area into an area small enough for focus, but large enough for an interesting analysis. There must be a transition connecting background information to goals and objectives.

6.2.3 STATE THE SPECIFIC GOALS OR OBJECTIVES

Stating specific goals or objectives is a critical part of the introduction. Readers will often skip ahead to the goals or objectives to see exactly what to expect throughout other sections of the writing. The goals or objectives can often influence the organization of other parts of the writing, and should identify important products resulting from analysis.

The goals for a project define real questions needing answers. The introduction should present information that reveals the question with evidence that there could be two or more plausible answers. The goal is to resolve which of the plausible answers is right, or to reveal a previously unconsidered answer. Authors should look carefully at an objective, and determine that it is real and resolvable.

For example, a strong question or goal is, "To compare the yields of economically important crop plants to ozone air pollution." On the one hand, ozone is toxic to biological cells, and therefore there is no reason to believe that cells of common crop plants will differ in ozone sensitivity. If so, agronomists only need to understand the impacts of ozone on one crop, and, by inference, understand the impacts on all crops. On the other hand, crop plants may have physiologies that are sufficiently different to impart important differences in ozone sensitivity. If so, research to discover the impacts of ozone on agriculture must include experiments on all crop species.

A related, weak question or goal is, "To show ozone air pollution affects crop production." The statement is an example of a straw man because the author (and most readers) already know the answer to the question.

Authors should also avoid normative science that results when creating objectives or questions that have a presumed, unstated answer. Put another way, authors should not use the pretext of an analytical paper to prove their point. For example, an author may have a deep interest and commitment in producing electricity from solar panels rather than from coal fired power plants. An objective or goal such as, "To show the benefits of shifting from electricity produced from coal to solar panels" puts the author in the position of using data for a predetermined purpose, and is therefore not an objective, scientific question.

The simplest approach to presenting the specific goals or objectives is to present an enumerated list, as shown in the example, below. Typically, there should be at least two goals or objectives, but probably not more than five. If there is a single goal, the goal may be simply too broad. If there are more than five goals, the paper will typically be larger than most readers will manage. Number the goals or objectives in logical order, with results or products of the first goal being connected to subsequent goals.

"The objectives of this paper are:

1. identify all the bird species affected by the Valdez oil spill;

2. prioritize the bird species from most affected to least affected by the oil spill;

3. determine the costs of rescue for the bird species most sensitive to the oil spill."

Start by simply stating each goal or objective with a simple, declarative sentence. Simply stating each goal can then lead to further development with several more sentences that further explain the importance or relevance of a goal. For example:

"The goal of this paper is to better understand the impacts of the Valdez oil spill on sea birds, and to evaluate the cost effectiveness of bird rescue efforts. Specific goals, listed below, focus on identification of bird species, ranking species on the basis of sensitivity to oil, and assessing costs for bird rescue.

1. **Identify all the bird species affected by the Valdez oil spill.** The sea bird rescue effort following the spill was extensive, involving the work of many people from a number of agencies. Numerous state and federal agencies, non-governmental agencies (NGO's), and volunteer groups provide sea bird rescue data including species lists. The final list of species of rescued sea birds will include a synthesis of data from all bird rescue groups.

2. **Rank bird species from most affected to least affected by the oil spill.** Bird species range in sensitivity to the spilled oil because of differences in exposure related to their feeding, roosting, swimming, flying, and other behaviors. Also, birds differ in their intrinsic capacity to shed oil, clean themselves, and resist toxic affects of ingested oil. Ranking bird species sensitivity to oil provides insights for prioritizing bird rescue efforts.

3. **Determine costs of rescue for individuals of each bird species.** Costs for rescue of individual birds differs for each species because of differences in the required amount of hand cleaning, post cleaning maintenance, or needed veterinarian care. Knowing the cost per individual bird may correlate with sensitivity to the oil spill, or with listings as endangered species. Understanding the rescue cost per individual will also help plan and budget for rescuing birds following oil spills in the future.

The examples above show just how carefully writers must choose the wording for their goals. Being clear with the goals means presenting a precise statement of what you will be doing, and why your analysis is important. The wording of the goals will invariably change as writers think through the project, and their understanding of the work evolves while working with the content.

6.3   APPROACH AND METHODS

The approach and methods section of writing should explain how the writer will accomplish the specific goals identified in the introduction. The approach and methods section of a paper should provide a complete explanation allowing the reader to understand the thought process behind the work, and to reproduce experiments or calculations. There could be many ways to accomplish goals specified in the introduction, and the author should justify chosen approaches.

The content of the approach/methods section of the paper should feature the reasoning and thought patterns necessary to accomplish the goals and objectives. The author must explain the thought patterns, decision trees, equations, statistical tests and other forms of reasoning necessary to accomplish the work.

6.3.1   APPROACH

A specific objective of the writing might be "To determine public opinion about global warming." Yet, there are many ways to determine public opinion, and the author should explain the reasoning for choosing one approach over another.

One approach might be to find a completed survey or poll. An alternative approach might be to create a new survey. A reason for using a completed survey is that data already exist and can quickly be put to use. A reason for conducting a new survey is to get current opinion from a targeted group. In either case, the writer should explain their approach for determining public opinion.

Presenting a rationale for determining public opinion on global warming provides the reader with important information. One reader might be interested in completed polls because they are likely to be rigorous, extensive, and recognized by others. Another reader might be more interested in a new survey designed to compare current opinions of targeted groups.

Writers can help define an approach when they describe their choice of field site, their rationale for choosing a study organism, or the type of mathematical model they will use. Regardless of how the analysis will proceed, writers will do well to outline the approach so readers know what to expect in subsequent sections of the article.

6.3.2   METHODS

Analytical writing often includes a section outlining details of methods used for experiments and data analysis. Methods used are important to readers because all methods have limitations, and some methods are better than others. Importantly, the use of multiple methods to get at the same point can be more convincing than the use of a single method. In addition, readers may want to know the details of techniques and methods to either repeat the analysis or experiment or to accomplish a new purpose.

The methods section of analytical writing is the "cook book" section, is straightforward, and presents a discussion justifying the choice of a specific method for the analysis, implementing the method, and explaining the sources and use of data. Writers should describe methods in sufficient detail so readers can understand the techniques, sampling protocols, statistical treatments, and methods for data processing. The methods might include step-by-step procedures for a laboratory experiment, or outline details of procedures followed in field studies. An engineer might provide specifics about parameters for validating mathematical models.

6.3.3 RESULTS

One approach to organizing the writing in the results section is to use subheadings that are roughly similar to the objectives or goals listed in the introduction section of the manuscript. The results section should present a story explaining the analysis and its outcome. Much of the writing in the results section is often centered on data, figures, graphs, maps, and photographs. The organization of the writing should reveal the information in a logical order, with ideas building systematically to complete the story.

The level of writing in the introduction is also important. Writers often tend to explain the tables or figures in excruciating detail, when that information is obvious to the reader who just glances at the graphic. While the information in tables and graphics needs some explanation, the story of the results should include an explanation of what the results mean, and how the results compare with similar or related work from others.

6.3.4 DISCUSSION

The discussion section allows the author the chance to explain the importance and significance of the results. The objectives and goals stated in the introduction can form subheading themes for the discussion section.

The writing in the discussion section should not retell the story found in the results section. Rather, the focus should be on the importance of the results, including such topics as the following.

1. **Are the results significant?** Perhaps the results lead to a major paradigm shift; if so, trumpet the news in the discussion section. Alternatively, a writer may explain the results represent an incremental advancement of an established idea.

2. **What are the limitations of the analysis?** Writers can explain whether the results of the analysis apply only to a specific situation, or whether the results logically extend to

other situations, locations, organisms, ecosystems, or points in time. The author should explain the extent to which the results apply to other situations.

Since no analysis is ever totally complete or satisfactory, the author should identify the uncertainty of the analysis completed for the project. Writers can explain whether the statistical power of the analysis is either weak or strong, and thoughts about how to reduce statistical uncertainty in future work.

All experiments have artifacts or unavoidable imperfections in the control treatment, the experimental condition, replication, or other aspects of analytical approaches. The author can partially defuse potential criticism by acknowledging the shortcomings of experimental design. Explaining the shortcomings does not necessarily weaken the conclusions drawn from the work, but ensures the results are put in proper context. Acknowledging shortcomings puts the writer in a position of authority and control of the interpretation of the results.

Analysis usually requires making some assumptions about data, models, locations, and many other features. The discussion section is the place to outline which assumptions in the analysis are significant, and the implications if the assumptions do not hold true. For example, models that simulate biological or physical processes operate with data taken from the published results of other workers. By using data from other sources, authors assume the data are appropriate for the model. Writers should explain whether the violation of the assumptions has either a large or small impact on the interpretation of the analysis.

3. **What are other approaches?** After completing a project involving analytical thinking and writing, the authors often come to understand the values of other approaches to create a more complete picture. Writers should identify which of these alternative approaches are most cost effective, complementary to existing work, and available for use. Similarly, writers should explain why some alternative approaches may seem to be useful, but are not.

4. **What are the next questions?** A thorough analysis of an issue will provide useful information for a decision or recommendation. Importantly, an analysis should provide insight for confronting the next set of issues. The discussion section can clearly identify the next issues in order of importance, ease of accomplishment, or likely impact. Such guidance is a useful outcome of analytical writing and can set the stage for the next project that builds sequentially, in a logical order.

6.4   BACK MATTER

The back matter of a manuscript allows the writer to finish the project by acknowledging those who contributed to the project, giving attribution to those cited in the text, and appending material that is too bulky in format and content to include in the main body of writing. Proper attention to back matter can provide polish for a project, and show readers a fully complete writing effort.

6.4.1   ACKNOWLEDGEMENTS

The acknowledgements section of a manuscript is typically a single, short paragraph that comes at the end of the discussion section. The content of the acknowledgements is highly personal to the author(s), and provides the opportunity to thank those who in some way contributed important pieces to the project, but who do not warrant authorship. A graduate student writing a thesis might acknowledge their parents for their life-long support of learning. A professor writing a peer-reviewed journal article might acknowledge staff members at a research center that helped with access to a remote field research site. An engineer or consultant might acknowledge those who helped draft figures for a book chapter. Grant proposals usually do not have an acknowledgements section.

6.4.2   REFERENCES

Analytical writing always includes a reference list, unless the author presents all, full references as footnotes in the text, which is a rare practice. Basic information for each specific reference should allow the reader the ability to find and read the same articles read by the author. The reference list is built from the citations in the narrative or graphical elements giving attribution for statements of fact. Each citation takes the reader to a specific reference in the reference list.

Authors can cite and list any form of communication provided readers can find them. For example, authors can cite and refer to films and movies, taped interviews, and websites (although not often). The key is to only cite and add to a reference list those items that can be found again, and to provide all the information necessary to re-find the original material.

With few exceptions, websites are not suitable as citations in the narrative or as references in the back matter. Authors citing and referencing websites have no assurance the information will exist at the website when a future reader tries to re-find the material in the reference. However, websites are an important source of references to articles, data, and other resources for writers.

Authors must see the piece of information for themselves before citing it and putting it into the reference list. Using citations and listing references on the basis of hearsay, or taking references from other sources without seeing the cited and referenced article, is not an acceptable writing practice.

The format of the references in the reference list may be specified by a journal or by funding agencies, and academic institutions may specify reference formats for theses. More details about formatting references appear in Chapter 8. In most cases, authors can choose a reference format, but all references in the list should be in the same format. There are three common, basic types of written material cited in analytical writing, and presented in the reference list: 1) journal articles, 2) books, and 3) chapters in edited books.

An example of a typical reference for a journal article containing all the required information is:

- Black, R.A. and R.N. Mack. 1986. Mount St. Helens ash: recreating its effects on the steppe environment and ecophysiology. Ecology 67:1289-1302.

Information contained in the journal article reference above includes:

1. the name of the author(s), R.A. Black and R.N. Mack;

2. the year the article was published, 1986;

3. the title of the article, Mount St. Helens ash: recreating its effects on the steppe environment and ecophysiology;

4. the journal that published the article, *Ecology*;

5. the volume and page numbers where the article can be found, 67:1289-1302.

References to books should include the information above, along with the publisher and the number of pages. An example of a reference for a book is:

- Nobel, P.S. 1991. *Physiochemical and Environmental Plant Physiology*. Academic Press. 635 p.

Edited books consist of a collection of chapters, often from different authors. When including chapters in edited books in a reference list, include the title and author(s) of the chapter and the title and editors of the book. An example of a reference to a chapter in an edited book is:

- Ammerman, J.W. 1993. Microbial cycling of inorganic and organic phosphorus in the water column. Pp. 649-660. In: Kemp, P.F., B.F. Sherr, E.B. Sherr, and J.J. Cole (eds.), *Handbook of Methods in Aquatic Microbial Ecology*, Lewis Publishers, Boca Raton, FL.

6.4.3 APPENDICES

In general, most analytical writing will not include an appendix. However, analytical writing may include material too lengthy to include in the main body of the writing, and is therefore appended to the back of the article. Some funding agencies may not allow proposals with appendices, but their use is typically left to the discretion of the writer. Appendices may include long lists of supporting documentation, raw data, maps, computer code, or other material either used in the analysis or that helps convey the author's intent. Cite appendices numerically in the narrative in the order presented. The appendices should not exceed the length of the text.

CHAPTER 7

The Writing Process

7.1 BUILD A BLUE PRINT FOR THE PROJECT

Before starting a writing project, the author should fully considering the time and other resources needed to successfully complete the work. Underestimating the investment to complete the project is common, especially for inexperienced writers. One approach to fully understand the effort needed for a project is to build a set of plans, or a blue print, that is similar to that developed by an architect designing a house. In this case, however, the writer is building a manuscript, so the blue print takes a unique form. Combining a time line with an estimated financial budget for the writing assignment provides an excellent starting point for a successful project.

7.1.1 TIME LINE

Writers must develop a timeline as a part of their writing blue print. One approach for developing a timeline is to work backward from the submission date or dead line. For example, if a deadline for a term paper, grant proposal, business plan is June 1, and the current date is January 1, a time line might be as follows:

Writing Project Time Line

- June 1: Submit final writing

- May 15 – May 31: Final edits, changes, and revisions

- May 1 – May 15: Revise manuscript and polish text

- April 21 - April 30: Get comments from outside readers

- April 10 – April 20: Polish text, tables, and figures

- March 29 – April 10: Finish second draft

- March 12 – March 28: Finish first draft with rough text, tables, and figures

- March 1 – March 12: Make tables, figures, maps, and process color photographs

- February 12 – March 1: Collect and analyze data

- February 11: Complete interviews with experts

- February 9 – 10: Revise outline adding detailed, analytical approach

- February 8: Commit to specific questions, goals, or hypotheses

- January 28 – February 7: Continue reading collected materials

- January 27: Revise outline to provide writing categories and structure

- January 26: Schedule interviews with experts in the field

- January 15 – 25: Collect maps and take color, digital photographs

- January 5 – 15: Begin library work and background reading

- January 2 – 5: Begin writing an outline and develop a list of ideas

- January 1: Decide general topic

7.1.2 FINANCIAL COSTS

Writing projects cost money. For employed professionals, costs to the employer for all involved include salaries and benefits, travel costs, and costs for space, utilities, computer and graphics facilities, and postage. Employers typically pay the costs of producing and reviewing written documents, and may provide a budget to do so.

High school and undergraduate students at colleges and universities typically pay for the costs for their writing projects. However, some schools may provide small amounts of funding to support student projects. Costs for writing projects can include:

- Making copies of text, documents, reports, maps, and other materials;

- Travel costs to gather information;

- Costs for books and reports needed for resources;

- Costs of collecting and processing samples;

- Costs of computer analysis;

- Costs of color photography;

- Costs for producing and distributing surveys;

- Costs for attending conferences or workshops;

- Costs for producing your written project.

7.2   OUTLINING

The process of analytical writing should always begin by building an outline as part of the project blueprint. The process of outlining is the first meeting place for analytical writing and thinking. Outlining a document is therefore a complex, iterative process resulting in refinement.

The context for a writing project is a critical part of the outlining process. A high school or college student will want to develop an outline for a term paper. A graduate student will want to develop an outline for a thesis. A professional scientist, engineer, or businessman will want to develop an outline for a professional publication, a construction proposal, or a business plan. Single author projects need outlines, as do team-based projects where outlining is a shared activity. Outlines are not only an essential starting point for analytical writing, but also for oral presentations, speeches, and seminars.

Writers use outlines to ensure ideas develop in logical order. In general, writing should start with a compelling reason for the project, explain the importance of the work, show the approaches and methods, present findings and conclusions, and summarize the importance of the new information. Outlines help establish the most logical way to present the large body of material and thinking that will result in a significant writing project. The idea is to ensure that the reader understands the document, the thought process underlying the writing process, and avoid gaps in reasoning or logic.

Writers also use outlines to ensure efficient use of writing time. By building an outline, authors avoid gathering irrelevant information or data, or conducting analyses and writing text that will not fit into the document. By sorting through the ideas and structure of the writing in an outline form, authors focus on working with ideas and subject matter that are certain to have a place in the manuscript.

An outline defines the size of the project. Specifying the length of the sections of the writing gives scope and scale for the writing in each section, and breaks the writing into manageable pieces. Authors should put estimated page lengths on each part of the outline to understand and visualize the size of the document.

7.2.1   EXPLORATORY OUTLINE

The first phase of the outlining process is exploratory. The outline can begin with a title containing a few key words that do not connect or make a cohesive statement. Follow the title with a list of relevant ideas that come to mind. At the beginning of the outlining activity, the list of ideas relates to the words in the title, but are in no particular order.

A high school student might choose to write a term paper on salmon fishing. An example of an initial outline is:

Initial Outline

Title: **Fishing for Salmon**

Salmon species

Commercial fishing

Alaska, Canada, lower 48 states

Importance of other countries, e.g. Japan, Russia,

Sport fishing in Alaska, US

Fly fishing

Endangered species issues

7.2.2   OUTLINE MODIFICATION

Following the process of writing the initial outline, the student must read about topics in the outline. In some cases, the author may already be familiar with some of the topics, but must gather additional understanding by reading and talking with experts, to advance understanding. In other cases, the author may know nothing about a topic in the outline list, and must start by reading basic information. Regardless of whether the author is familiar with the topic, or not, all authors must go through a phase of gathering and assimilating information.

After outlining and reading, the high school student improves their understanding of the material, and understands how the topics of the term paper create a sequence for presenting ideas and information. The thinking leads to revisions and more sophisticated outlining. The modified outline now includes an organization of main ideas with subdivisions. The outline gains focus and shifts from describing fishing for salmon to include elements of analytical thinking that lead to useful ideas. In this example, the author will attempt to estimate the impacts of fishing on salmon populations.

The revised outline might appear:

Modified Outline I

Title: **Impacts of Salmon Fishing**

Existing salmon stocks

Salmon populations in Alaska, Canada, the lower 48 states

Salmon populations in Japan and Russia

Endangered salmon species and runs

Harvests from fishing

Commercial salmon fishing

Sport fishing in Alaska and the lower 48 states

Fly fishing

Predicting salmon populations

7.2.3   FUTHER REVISING

The modified outline has more focus than the original, but the author is still not ready to write. The student correctly concludes that there is more background work necessary to form a rigorous, analytical framework. But the interaction between analytical writing and thinking is underway.

The last exercise listed, "Predicting salmon populations," seems large and difficult. There are salmon throughout the Pacific Ocean, with some runs endangered, and some runs teeming with fish. The author decides to make predictions for specific salmon runs in a specific geographic region.

Since the student is a fly fisher in Alaska, and his father is a commercial salmon fisherman, the author really wants to know if fly fishing has adverse effects on commercial salmon harvests. To evaluate the effects of fly fishing on commercial harvests, the author considers two approaches.

Population models predict how salmon runs are affected by "predation" by bears. The models allow biologists to simulate bears and to mathematically remove artificial numbers of male and female fish from runs that spawn in freshwater streams. The student thinks that fish caught by fly fishing are similar to fish lost to bears, and can simply be considered a form of predation. The model will calculate whether or not the effects of those parental losses, from predators, affects the number of adult fish that can be caught three years later when mature fish return to the stream to spawn.

However, the population models are large, complex, and have assumptions about survival of eggs, smolts, and adult salmon at sea. So, the author seeks a more simple approach looking at the impacts of fly fishing on salmon in Alaska. The new idea is to determine if the timing of increasing pressure from fly fishing correlates with changes in commercial fishing harvests.

Data show that fly-fishing activity throughout all of Alaska increased 10 fold between 1960 and 2000. Also, commercial salmon harvests changed whether fly-fishing pressure was high or low. So the author further modifies the outline:

Modified Outline II

Title: **Impacts of Fly Fishing on Commercial Salmon Harvests in Alaska**

Existing Salmon Stocks in Alaska

The importance of salmon in Alaska

Commercial salmon species present

Known population trends

Commercial harvest trends

Potential threats to commercial salmon fishing

Environmental change and loss of habitat

Predation by mammals

Fishing pressures

The need to ensure sustainable salmon populations

Commercial values

Sport fishing industry

Cultural and subsistence values for Native Americans

Approaches for predicting impacts of fly fishing

Predictive models simulating population changes

Correlations between trends in fly fishing and commercial harvests

Correlations show no detectable fly-fishing impacts on commercial harvests

7.2.4   OUTLINE FOR MANUSCRIPT DRAFT

Writing is again clarifying thought. The author sees how the manuscript will take on a specific, analytical question, and will produce useful information. Yet, the thinking is not quite far enough along to begin writing the paper, and more conceptual work on the outline is necessary.

There are two gaps in thinking. The ideas in the first part of the outline do not lead the reader to understand the importance of making a prediction of fly fishing impacts on commercial harvests. In addition, the last item brings out the concept of correlations but just hangs and has no connections.

Finally, the outline, above, has main divisions and subdivisions, but needs more formal numbering to be emphatic and clear. Now the author is thinking about the figures, tables, front matter, and back matter for the paper. The next outline revision results in a form that provides the framework for the first draft of the manuscript:

Modified Outline III

Title: **Predict Impacts of Fly Fishing on Commercial Salmon Harvests in Alaska**

i. Title Page (1 Page)

ii. Table of Contents (1 Page)

1. Introduction (5 pages)

1. Background & Rationale: Existing Salmon Stocks in Alaska

1.1. The importance of salmon in Alaska

1.1.1. Commercial salmon species present

1.1.2. Known population trends

1.1.3. Commercial harvest trends

1.2. Potential threats to commercial salmon fishing

1.2.1. Environmental change and loss of habitat

1.2.2. Predation by mammals

1.2.3. Fly-fishing pressures

1.3. The need to predict salmon populations

1.3.1. Ensure sustainable commercial values

1.3.2. Ensure sustainable sport fishing industry

1.3.3. Ensure culture & subsistence for Native Americans

1.4. Objectives: Predictions

1.4.1. Impacts of bears on salmon runs

1.4.2. Impacts of fly fishing on salmon runs

2. Approaches for predicting impacts of fly fishing (3 pages)

2.1. Predictive models simulating population changes

2.2. Correlations between trends in fly fishing and commercial harvests

3. Results: Impacts of bears and fly fishing on salmon harvests (5 pages)

3.1. The models of predation

3.2. The importance of the correlations

3.3. Assumptions in the analysis by correlation

3.4. Limitations of the analysis

3.5. The need to use alternative analytical techniques

4. Discussion: Significance of fly-fishing impacts on salmon harvests (3 pages)

4.1. The importance of predation by bears (ecology)

4.2. The importance of fly fishing (economy & recreation)

4.3. Assumptions in the modeling

4.4. Further analysis

5. Acknowledgements

6. References (2 pages)

7. List of tables (2 pages)

    7.1. Map of Alaska salmon runs

    7.2. Salmon species list

8. List of figures (4 pages)

    8.1. Salmon population sizes in Alaska

    8.2. Trends in commercial harvests

    8.3. Trends in fly fishing activity

    8.4. Correlation of commercial fishing vs fly fishing

The author now has an outline that contains the elements of front matter, an introduction, approach, results, discussion, acknowledgements, and reference list. In addition, there is now a list of tables and figures that will be a part of the report. As a final touch, the author estimates and notes the page length for each major section, thinking carefully how each section of the paper will fit in context and size with the whole manuscript. The author has an outline for a 26 page paper, which is appropriate for the assignment.

7.2.5 MORE REVISIONS IN THE OUTLINE MAY OCCUR

The author happy with the outlining activity. The outline shows general information leading to objectives, a clear focus and objectives, an understanding of the analytical approach, and the content of the front and back matter. The author may use the outline, unaltered, to the completion of the project.

However, as work progresses, it is possible that more revisions in the outline will be necessary. For example, the author may decide to include a specific case study that documents fly fishing and commercial fishing on a specific river, and further support the conclusion that fly fishing has no impact on commercial fishing harvests. If so, the author should look at the new case study carefully, return to the outline, and revise the outline to integrate the case study into the manuscript. Alternatively, the author may decide that the issue of predation by bears is peripheral to the main points and not relevant. Revising the outline is the only way to ensure that addition or deletion of material originally in the outline does not result in gaps in logic, and that the manuscript remains coherent.

7.3 FIRST DRAFT

The process of writing the first draft is a critical phase of the project. The goal is simple; get some words in place that follow the outline. The first draft is a written document and is often a roughly hewn manuscript with typographical errors, poor sentence structure, and inconsistent formats for margins, lists, and headings. Regardless of how the manuscript appears, the process of writing the first draft engages the author, and is a highly rewarding experience. A complete first draft should include the front matter, narrative, back matter, figures, tables, references, acknowledgments, and appendices.

7.3.1   START WRITING

The writing process should begin immediately after finishing the outline. If too much time lags between the formation of ideas in the outline and writing, the loss of momentum for the project can severely affect the quality of the final product. When a lag occurs between outlining and writing, feelings of guilt develop and the time fills with other projects, issues. More serious problems develop when the writer moves the project out of mind, and then stops active thinking about the manuscript. Do not let the writing fall victim to procrastination that eventually leads to a hurried, unsatisfactory project.

7.3.2   USE THE OUTLINE

The outline provides the framework for the first draft that is parallel in structure to the outline. Use the literal form of the outline to establish sections and subsections of the paper. By using the outline for its main headings, subheadings, and subsections, the writer divides the text in an organized fashion and is able to stay oriented in the writing process. In the example of the outline, above, consider the following main headings, subheadings, and subsections. The formatting for organization chosen here is only meant to be one example of many authors can use:

Title: **Predicting Impacts of Fly Fishing on Commercial Salmon Harvests in Alaska**

    i. Title Page (1 page)

    ii. Table of Contents (1 page)

    1. Introduction (5 pages)

    1. Background & Rationale: Existing Salmon Stocks in Alaska

        1.1. The importance of salmon in Alaska

            1.1.1. Commercial salmon species present

            1.1.2. Known population trends

            1.1.3. Commercial harvest trends

        1.2. Potential threats to commercial salmon fishing

            1.2.1. Environmental change and loss of habitat

            1.2.2. Predation by mammals

            1.2.3. Fly-Fishing pressures

 1.3.   The need to predict salmon populations

  1.3.1. Ensure sustainable commercial values

  1.3.2. Ensure sustainable sport fishing industry

  1.3.3. Ensure culture & subsistence for Native Americans

 1.4. Objectives/Goals: Predictions of predation on salmon runs

  1.4.1. Impacts of bears on salmon runs

  1.4.2. Impacts of fly fishing on salmon runs

The first draft should include a cover page, followed by a page that is the table of contents. The two pages leading to the narrative have small Roman numerals for page numbers.

The first, main section of the writing is numbered and presented as

1. <u>Existing Salmon Stocks in Alaska</u>

Text from the author explaining the importance of existing salmon stocks in Alaska appears here. For the heading, place the heading number with one digit, in underlined bold at the left margin, hanging alone in a line. Capitalize the first letters of each word except prepositions and conjunctions, and create a line space between the main heading and the text.

Following narrative in Section 1 is the first subheading

 1.1. <u>The importance of salmon in Alaska.</u> Text on the importance of salmon in Alaska appears, as a subheading. Number the subheading with two digits, at the left margin, underlined, but not in bold. Capitalize only the first letter of the first word and proper nouns, and end the subheading with a period. Formatting can also include a line space between a subheading and a subsection.

Similarly, the section numbered "1.1" leads to three subsections, 1.1.1, 1.1.2, and 1.1.3. Each subsection more fully develops narrative specific to the subdivision heading, "The importance of salmon in Alaska."

  1.1.1. *Commercial salmon species present.* The text on existing, commercial salmon species appears here. The format for subsection headings includes indentation, three digit numbering, and italics. Capitalize only the first letter of the first word and proper nouns, and end the subsection heading with a period. The subsection heading uses a line space to provide separation from the next subsection.

  1.1.2. *Known population trends.* Another heading for a subsection, with the format the same as in 1.1.1., above.

1.1.3. Commercial harvest trends. Here is another heading for a subsection, with the format the same as in 1.1.1, above.

1.2. <u>Potential threats to commercial salmon fishing</u>. Another subheading, in the same format as the subheading for 1.1., above.

1.2.1. Environmental change and loss of habitat. Another subsection, presented as subsections, above.

1.2.2. Predation by mammals. Another subsection, presented as subsections, above.

1.2.3. Fishing pressures. Another subsection, presented as subsections, above.

Showing how writing the first draft in parallel structure with the outline stops here. The point is to use the outline to aid writing, to place writing in the right location to assure logical flow of ideas, to provide formatting and structure to the paper, and to focus the writing to specified numbers of pages.

7.3.3 THE INTRODUCTION MAY NOT BE A GOOD STARTING POINT FOR WRITING

Writing the introduction requires setting the stage and rationale for the project, and stating specific goals, objectives, or hypotheses. At the beginning of the effort to write the first draft the content of some of the introductory material, such as the scope and specific goals of the project, may be vague, ill formed, and difficult to capture in writing. If so, authors should consider beginning the first draft by writing a section of the paper that provides an easier starting point.

One idea is to begin writing the paper with the "approach" or the "methods" section that explains how the writer gathered and analyzed information. The explanation is typically straightforward, involves nuts and bolts, and is easier to write than other sections. The approach or the methods can also be a good writing starting point because it leads logically to subsequent sections that become progressively more difficult. Upon finishing the approach or methods section, the author can more easily explain the results in the next section of the narrative. The importance of the results follows next, in the discussion section. Writing the introduction, including a refinement of the objectives, is easier once the writer knows the importance of the results. In effect, the approach, results, and discussion set the stage for revealing the important, relevant background information, and leads to refined objectives and goals.

7.3.4 COMPLETE AN ENTIRE FIRST DRAFT

Once starting the first draft, keep working through the manuscript until an entire draft is complete. Avoid the temptation to revise sections as you go. Keep moving forward with new material, following the outline, until you complete a first draft of the entire document. Accept the fact that the writing is not perfect, and keep moving, writing new text.

7.4   REVISION

The process of writing is iterative as writing advances thinking. The outline is a tool to efficiently prepare a cohesive first draft that is the starting point for a series of revisions. In general, each draft of the paper results in smaller numbers of changes, less dramatic change, and a more polished piece of writing.

Proceeding from the first draft to the second draft involves thinking through the sequence for presenting ideas. The author should be sure the text introduces compelling issues, logically generates objectives, justifies gathering and analyzing information, and develops a comprehensive line of reasoning. In some cases, the outline may appear to be a strong, logical framework. However, writing the first draft may reveal gaps in information or logic not apparent in the outline, and represent areas that need attention and development. If so, writers must correct the outline, be sure it is still coherent, and then revise the text. When the process of revising the text includes changing the outline, the writer keeps the level of organization and vision necessary to begin creating the second draft.

The development of the second draft should not only reveal where new writing is necessary, but also must include editing to fix problems with spelling, word use, syntax, and grammar. Start by grooming the first draft, from the beginning to the end. Check first for organization and content, while also keeping an eye on the analytical elements to be sure the writing still accomplishes a stated purpose.

When the writer is sure that the main headings, subheadings, and subsections are in the right order and contain the right kind and amount of information or ideas, it is time to address other editing issues. The author might use software for Spell Check and Grammar Check to find and eliminate some errors. In addition, the writer should take this opportunity to replace specific words or clauses that are unclear or weak, and to rewrite sentences or paragraphs that are awkward, unclear, or confusing.

The revisions necessary for a second draft include evaluating the tables and figures. The author should look critically to be sure all the figures and tables are necessary, and that the narrative presents and discusses the graphical aids in the right locations. The tables and figures do not need to be in final form, but should have notes written on them to show necessary changes.

Revising the first draft also means carefully checking the citations in the text and the references listed at the end of the document. Each statement of fact in the text should have a citation. Each citation should be included in the reference list. Each item in the reference list should tie to a citation somewhere in the text. As the writer adds or deletes citations in the text, amendments to the reference are necessary. Similarly, if references are dropped or added to the reference list, appropriate changes must also occur with citations in the text. Synchronizing the citations and references is important as the manuscript evolves, and failure to attend to the citations and references can create problems with the final draft.

Critical readers should be able to understand and review a second draft of a manuscript. Such readers might include fellow students, others on the same writing team, colleagues, and peers. The best readers are those who have a critical eye, but will not directly evaluate the finished written product. Authors should explain to outside readers about the need for them to provide frank, insightful comments about your project, including the justification, the content and organization, the synthesis, the quality of analysis, and formatting issues. Authors should appreciate reviews and thank outside readers who take time to provide constructive criticism. Writers should not become defensive or try to further justify their work. Take pause, listen, and think carefully about what the outside readers suggest.

7.5   FINAL DRAFTS

When authors finish a second draft, receive comments from outside readers, and revise accordingly, the manuscript should be close to completion. The draft resulting from changes in response to comments from outside readers is probably not the final draft. Rather, if possible, writers should step away from the project to see it from a distance. When viewed from afar, authors should be sure the writing looks clear, sharp, and has the desired content.

Take the time to compare the paper to other papers that are similar in nature. For example, before submission students should be sure their thesis looks like other, accepted theses. Students submitting an article to an undergraduate research conference should read other articles submitted to the conference in the previous year. Entrepreneurs submitting a business plan to a bank should compare their plan with others supported by the bank.

After comparing documents, authors may want to make further revisions. In the process, writing may result in a third or fourth draft. If the document changes form dramatically from that read by outside readers, consider getting additional readers to comment on the latest version.

The final draft is complete when further work brings negligible improvements in the document. The author should feel proud to have their name on the cover page. Even so, once the final draft is submitted to a review panel, a graduate committee, an instructor, writers expect to get feedback from readers that will be complimentary about some features of writing, but will also offer suggestions. Journal editors, graduate committees, and instructors may ask for revisions before accepting a final form of the article.

CHAPTER 8

Construction

Writers often focus exclusively on producing the narrative for an analytical project or report, but there is much more to consider. Although the narrative should receive the main effort, there are other parts of the project, such as citations, references, and graphical elements that also warrant attention. Effective writers allow time to construct all the features of the written project.

8.1 NARRATIVE (SYNTAX AND GRAMMAR)

This handbook presents ideas for producing the narrative, or written text, that forms the body of a report. The focus on narrative will ensure the writer crafts a well-structured story, but the construction and syntax of the sentences and writing warrants constant attention. In the most general sense, syntax is the study of the rules of language. For the sake of developing analytical writing skill, syntax refers to creating sentences that are easily read and understood. Writers will constantly "Word Smith" or hammer on the syntax of problematic sentences to improve their structure. The "Top 10 Writing Tips" that follow can help guide "Word Smithing."

Authors should be aware that text viewed on a computer screen reads differently than does printed text. Writers should address syntax issues by occasionally printing the document and reading it.

8.2 CITATIONS

Citations link a statement of fact made in the text to a specific source of information provided in a reference list at the end of the narrative. Citations should provide the reader links to information that include books by single authors, books with multiple editors, journal articles, reports from the public and private sectors, movies, photographs, pieces of art, or any other documentation that is important as a statement of fact for the narrative.

Authors should have hard copy of documents before using the document as a cited source of information. Writers should never cite unseen documents or information, verbal information about a source, or use an unseen citation from another author. Writers must see the cited evidence for themselves. Similarly, do not cite articles that are not accessible to readers, or articles not available online or in libraries.

Writers are often uncertain if a statement of fact needs a citation. In general, citations are not necessary for common knowledge. For example, we all know that objects fall toward earth, that water freezes, and that birds fly, and citations for statements like these are not appropriate.

However, stating that magnetic north differs between the poles and equator, that salt changes the freezing point of water, or that some birds cannot fly would all warrant a citation.

The web can be an important source of information, but should rarely be cited. In many instances, the web provides convenient access to information in libraries and offices or agencies in the public and private sectors. Reputable websites include those providing access to information in libraries, professional journals, public agencies, and well-known professional groups and businesses.

Avoid citing websites for many reasons. The value of the citation and reference is that readers can check the basis for the statements of facts made by an author. Readers may use articles that are a decade or more old, and there is no way authors can ensure information taken from a website will exist a decade, or more, into the future. In fact, there is no way to ensure the website will exist far into the future. In addition, many eccentric groups spread misinformation on the web, so care must be taken to be sure the information gathered online is credible.

Websites are appropriate as sources of information in rare cases. For example, an author might write an article about websites. Even so, the document should include citations and references from hard copy documents.

There are a number of ways to cite statements of fact. For example, writers can put citations in parentheses (parenthetical elements) within or at the end of sentences. When using citations in parentheses, the citation typically includes the author(s) and date, and is linked to a list of references at the end of the document. Two examples of parenthetical citations in the text follow.

8.2.1   CITATION EXAMPLE #1

"The values of ecosystem services are notoriously difficult to estimate (Contanza, et. al., 1997), yet can be applied to the issues of carbon sequestration. Terrestrial ecosystems can be sources or sinks of carbon, and therefore contribute to or reduce the build-up of carbon dioxide in the atmosphere (Houghton, 1996). Tropical forests have considerable biomass and the potential to store and cycle large quantities of carbon (Brown and Lugo, 1990). Even so, the pressures to convert native forests to high-yield forests (Gladstone and Legid, 1990), and the variation in carbon content within and between geotropical forest species (Elias and Potvin, 2003) will challenge those interested in carbon sequestration in tropical ecosystems."

Note that in the example above, the citation can be placed in the middle or at the end of the sentence. In all cases, the citation is placed immediately following the statement of fact. One advantage to parenthetical citations is that the reader can immediately see the source of evidence. In many cases, informed readers will recognize the citation, and either accept, reject, or question the statement based on the citation.

An advantage of using author citations in the text is that writers simplify the process of adding and dropping specific citations. In general, a specific information source is only cited once in a document. When a specific source of information is cited twice the citations should be far from each other and should typically refer to different data or ideas found within the same information source.

The format of the information in the citation may be specified or the author may be free to choose a format used consistently throughout the narrative. There are many examples of citation formats, but writers must include an author and a date to link to a specific article in the reference list:

- Single author, with or without a comma between the author name and the date:

  ◦ (Houghton, 1996)

  ◦ (Houghton 1996)

- Two authors, and the date:

  ◦ (Brown and Lugo, 1990)

  ◦ (Brown & Lugo 1990)

- *Three authors, or more, and the date:*

  ◦ Costanza, et al., 1997

The example below shows how numbers can link factual statements in the narrative to a specific reference. More specifically, the number "1" appearing either as a superscript, or in parentheses, following a statement of fact in the narrative means the reader can go to the reference list at the end of the document, find the item in the list with the number "1," and see the information source for the statement.

8.2.2   CITATION EXAMPLE #2

"The values of ecosystem services are notoriously difficult to estimate,[1] yet can be applied to the issues of carbon sequestration. Terrestrial ecosystems can be sources or sinks of carbon, and therefore contribute to or reduce the build-up of carbon dioxide in the atmosphere.[2] Tropical forests have considerable biomass and the potential to store and cycle large quantities of carbon.[3] Even so, the pressures to convert native forests to high-yield forests,[4] and the variation in carbon content within and between geotropically forest species[5] will challenge those interested in carbon sequestration in tropical ecosystems."

Writers use a numbering system for linking statements of fact in the narrative with references to make the document easier to read. However, using a numbering system makes it difficult to keep track of references during the writing process as authors either add or delete citations and references. In addition, using numbering systems for citations inconveniences readers who must continually move between the narrative and the reference list to identify cited sources.

Footnotes can also document statements of fact. Footnotes appear with numbers in the narrative, with the numbers referring to a numbered footnote that appears at the bottom of the page. Footnotes might identify a specific journal article or book, but can also include documentation for conversations, events at meetings, or other forms of evidence important to the narrative. When using footnotes, there is no reference list at the end of the narrative because the information sources appear at the bottom of each page. Documents written with footnotes generally have few references. If there are large numbers of footnotes on a page, there is not much room for narrative, and the text can lose cohesion for the reader.

Writers can also document conversations, and other informal evidence, as the source of statements of fact. For example, a student might interview a professor, Dr. Jones, about melting of the polar ice cap. The student might learn during the interview that Dr. Jones has new data from a recent trip showing melting of polar ice is increasing the relative humidity of air over the North Pole. Although the data are not yet published, and if Dr. Jones agrees, the student can write:

"Melting of the polar ice cap is increasing the relative humidity of air over the North Pole (Personal communication, 2013, Dr. A. Jones, Department of Earth Sciences, National Climate Change University)." Such citations in the text do not need listing in the reference list.

8.3   REFERENCES

Writers usually present the list of reference at the end of the narrative. An important part of proof reading is cross checking citations and references. Each citation in the narrative, whether in parentheses or numbered, should lead to a specific document in the reference list. Each document in the reference list must have a citation in the text.

A reference list can include virtually any source of information available to readers. Information sources commonly found in reference lists include written reports that are accessible, journal articles, books, book chapters, newspapers, art, speeches, movies, and music. The only requirement is that the reference allows the reader to find the specific source of information the author used in a statement of fact.

As with citations, the format for the reference list may be specified for authors, and if not, writers can choose their own format. Such decisions should involve a group discussion if a writing team is involved. The key is to determine if there is a required format, and if so, to follow it. If not, establish a format, and be consistent with it.

Examples of formats used for reference lists are presented, below, with examples of nearly infinite combinations of punctuation, name arrangements, placement of publication date, and identification of page numbering in books and journal articles.

Examples of reference lists come from the sample narrative with citations, above. The reference list below is taken from the Citation Example #1, above.

List references alphabetically by the lead author's last name:

Brown, S. and A. E. Lugo. 1990. Tropical secondary forests. *Journal of Tropical Ecology* 6: 1-32.

Costanza, R., R. d'Arge, R. de Groot, S. Farber, M. Grasso, B. Hannon, K. Limburg, S. Naeem, R. O'Neill, J Paruelo, R. Raskin, P. Sutton, and M. van den Belt. 1997. The value of the world's ecosystem services and natural capital. *Nature* 387: 253–260.

Elias, M., and C. Potvin. 2003. Assessing inter- and intra-specific variation in trunk carbon concentration for 32 geotropically tree species. *Canadian Journal of Forest Resources* 33: 1039-1045.

Gladstone, W.T., and F.T. Legid. 1990. Reducing pressure on natural forest through high-yield forestry. *Forestry Ecology and Management* 35: 69-78.

Houghton, R.A. 1996. Converting terrestrial ecosystems from sources to sinks of carbon. *Ambio* 25: 267-225.

Authors must build a reference list even when using numbered citations in the narrative. The reference list that follows is made from Citation Example #2, above:

List numbered references by order in which they appear:

1. Costanza, R., R. d'Arge, R. de Groot, S. Farber, M. Grasso, B. Hannon, K. Limburg, S. Naeem, R. O'Neill, J Paruelo, R. Raskin, P. Sutton, and M. van den Belt. 1997. The value of the world's ecosystem services and natural capital. *Nature* 387: 253 – 260.

2. Houghton, R.A. 1996. Converting terrestrial ecosystems from sources to sinks of carbon. *Ambio* 25: 267-225.

3. Brown, S. and A. E. Lugo. 1990. Tropical secondary forests. *Journal of Tropical Ecology* 6: 1-32.

4. Gladstone, W.T., and F.T. Legid. 1990. Reducing pressure on natural forest through high-yield forestry. *Forestry Ecology and Management* 35: 69-78.

5. Elias, M., and C. Potvin. 2003. Assessing inter- and intra-specific variation in trunk carbon concentration for 32 geotropically tree species. *Canadian Journal of Forest Resources* 33: 1039-1045.

The format for each reference in the reference list is important to consider and should be consistent through the list. Examples of reference formatting include:

Use of commas or periods:

Brown, S. and A. E. Lugo, 1990, Tropical secondary forests, *Journal of Tropical Ecology*, 6, 1-32.

Brown, S. and A. E. Lugo. 1990. Tropical secondary forests. *Journal of Tropical Ecology* 6: 1-32.

Formats for authors list:

Houghton, R.A. 1996. Converting terrestrial ecosystems from sources to sinks of carbon. *Ambio* 25: 267-225.

Brown, S. and A. E. Lugo. 1990. Tropical secondary forests. *Journal of Tropical Ecology* 6: 1-32.

Costanza, R., R. d'Arge, R. de Groot, S. Farber, M. Grasso, B. Hannon, K. Limburg, S. Naeem, R. O'Neill, J Paruelo, R. Raskin, P. Sutton, and M. van den Belt. 1997. The value of the world's ecosystem services and natural capital. *Nature* 387: 253–260.

Costanza, R., *et. al.* 1977. The value of the world's ecosystem services and natural capital. *Nature* 387: 253–260.

Publication date follow authors or placed at end:

Brown, S. and A. E. Lugo. 1990. Tropical secondary forests. *Journal of Tropical Ecology* 6: 1-32.

Brown, S. and A. E. Lugo. Tropical secondary forests. *Journal of Tropical Ecology* 6: 1-32. 1990.

Journal articles, books, book chapters, reports, organizational authors, websites:

Brown, S. and A. E. Lugo. 1990. Tropical secondary forests. *Journal of Tropical Ecology* 6: 1-32.

Larcher, W. 2003. *Physiological Plant Ecology: Ecophysiology and Stress Physiology of Functional Groups*. Springer, Berlin, Germany.

Madrid, S., and F. Chapela. 2003. Annex III: Certification in Mexico: The cases of Durango and Oaxaca. Pp. 1-2. In: Molnar, A. (ed.). 2003. *Forest Certification and Communitieis: Looking forward to the Next Decade*. Pub. Forest Trends, Washington, D.C.

Chemonics International, Inc. 2003. Community forestry management in the Maya Biosphere Reserve: Close to financial self-sufficiency? *Guatemala BIOFOR IQC Task Order 815.*

Stern, M. 2004. As far as I can throw 'em: Expanding the paradigm for park/people studies beyond economic rationality. Paper presented at conference, *People in Parks: Beyond the Debate. Annual Conference of the International Society of Tropical Foresters*, Yale School of Forestry and Environmental Studies chapter, New Haven, Connecticut.

WRM (World Rainforest Movement). 2003. Ecuador: Mangroves and shrimp farming companies. `http://www.wrm.org.uy/bulletin/51/Ecuador.html`

8.4   GRAPHICAL ELEMENTS: TABLES AND FIGURES

The graphical elements of analytical writing typically include tables and figures, but could also include maps, photographs, artwork, cartoons, or any other images. Tables and figures provide writers with exciting ways to bring information into a piece of analytical writing. Considerable thought is necessary when creating tables and figures because construction is time-consuming. Even with the use of software designed specifically for creating tables and figures, designing and building these elements requires patience. Although the initial plots may look satisfactory, it invariably takes time to work through all the details necessary to properly label, color, and size graphical elements.

8.4.1   SOURCES OF DATA FOR TABLES AND FIGURES

Authors can take tables and figures taken from other sources and include them in new writing. To avoid plagiarism, writers must cite the source of the graphical element in either the table heading or figure legend, and further identify the original source in the reference list.

Authors can also take data from several sources and create their own, original table or figure. Here, the author must cite the original data sources in the table heading or figure legend, and list the original sources in the reference list.

Of course, writers can create their own tables and figures from data and information they discover in original experiments, surveys, and other activities that produce new information. However, analytical writing does not require authors to do experimental research, but does require original thought and applications for existing information.

8.4.2   FORMATTING

Figures can be maps, graphs, pictures, drawing, photographs, or any other visual aid that helps convey the writer's message. Figures can be in color or black and white, two dimensional, or multi-dimensional, and can be in sections or pieces. When a single figure contains several elements that are disconnected, each element has a label, often a letter, so the author can separately discuss each segment. For example, a figure might contain three separate graphs labeled Figure 3A, 3B, and 3C, and identified as such in a single figure legend for Figure 3.

In some cases, authors may find requirements for formatting tables and figures, or limits on the number of them. For example, the graduate school may have specific guidelines for formatting tables and for graphs included in a thesis. Journals may also have format guidelines for tables and graphs. In other cases, authors may be free to choose how to present tables and figures. If so, the author should decide on a format for tables, decide on a format for figures, and these formats should remain consistent throughout the document.

Because publishing graphical elements consumes time and page space, and is expensive, writers use only essential tables and figures, keeping the number to a minimum. If the writer has doubts about including a table or figure, they exclude it. If the writer has trouble building a paragraph around a table or figure, they exclude it. If the table or figure is not substantial, is not well produced, looks weak, has weak data, or contains information for which the source is not known, exclude it. Readers will have an easier time if documents contain a mix of tables and figures, with figures containing a mix of forms including bar graphs, pie charts, and histograms.

Tables and figures should never show the same data. For example, if the writer presents a graph showing how atmospheric concentrations of carbon dioxide in the atmosphere have increased since 1960, the same data should not be presented as a table in the same document.

Tables, graphs, maps, and other graphical aids are best when the lay out is simple and easy for readers to comprehend. Readers, and editors, prefer tables that do not have too many rows or columns. Similarly, graphs should not have too many lines, colors, or symbols that are difficult to distinguish. Tables or graphs, photographs, and maps that are too small for easy viewing will not invite readers to look at them. A rule of thumb is that tables and figures should not be more than half a page in size, and never more than one page in size.

Authors should present tables and figures, numbered in the order in the text. Tables are numbered sequentially. For example, the first table presented is Table 1. The second table presented is Table 2, and so on. The first figure presented, whether a graph, map, photograph, or some other graphical element is Figure 1.

8.4.3 CITATION

Analytical writing includes some conventions for citing figures and tables. One common approach is to introduce a graphical element as a parenthetical element in the topic sentence of a paragraph. The effect is for the subject matter of the sentence to be the prime focus, with evidence from the graphical element cited as the evidence for the statement. The citation for the table and figure appears in exactly the same way as for a citation to a specific reference that also establishes a fact.

The four examples, below, show how to cite Figure 1, a graph of atmospheric carbon dioxide concentrations. The same formats apply to citing tables.

Example #1

"Carbon dioxide concentrations in the atmosphere increased during the past four decades (Figure 1)."

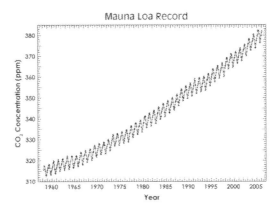

Figure 1. The change in atmospheric carbon dioxide concentration measured at Mauna Loa, Hawaii, from 1958–2005 (Tans and Keeling, 2013).

Figure 1 is an example of the rare instance when citing and referencing a website is appropriate. The NOAA website, staffed by Tans and Keeling, provides the most authoratative data for atmospheric CO_2 concentrations. The website is long established, annotated with historic record keeping for data and reports, and gives specific instructions to those writing citations and references.

In the sentence, Example #1 above, the idea that carbon dioxide concentrations are increasing is the subject and sets the stage for further discussion of the importance of understanding changes in carbon dioxide concentrations, as follows:

Building example #1 into a paragraph

"Carbon dioxide concentrations in the atmosphere increased during the past four decades (Figure 1). Carbon dioxide concentrations are increasing in the atmosphere because of burning fossil fuels and land use practices. Since carbon dioxide is a greenhouse gas, increasing atmospheric carbon dioxide concentrations cause global climate change including changes in air temperatures, precipitation, wind speed, and storm frequency. Such climate change brings uncertainty to those charged with managing natural resources including forests, agricultural lands, and wild lands."

Another form of the sentence, and a form to avoid, has the undesired affect of making Figure 1 the subject of the sentence:

Example #2

"Figure 1 shows carbon dioxide concentrations in the atmosphere increased during the past four decades."

In example #2, Figure 1 is unfortunately the subject of the sentence. The format of Example #2, making the figure the subject, should only be done when the subject of the sentence, and the message of the paragraph, is really about the figure, as below:

Building example #2 into a paragraph

"Figure 1 shows carbon dioxide concentrations in the atmosphere increased during the past four decades. The figure is an excellent example of how to plot progressive change in an environmental feature. Similar approaches could show how concentrations of mercury, nitrogen, and sulfur have changed in air, soils, streams, and lakes."

8.4.4   FIGURES AND TABLES IN THE CONTEXT OF A PARAGRAPH

Authors typically cite tables and figures only once in a piece of analytical writing. Make the citation at the end of the topic sentence of a paragraph, develop the paragraph with the ideas central to the figure or table, and be done with it. Then, move on to the next paragraph that may begin with a topic sentence citing the next table or figure. Citing a graphical element more than once, requires placing citations far apart in the text and that refer to different features.

Only one figure or table should typically be cited in a paragraph. If authors cannot develop the three to five sentences about a graphical element, exclude it from the paper.

8.4.5   CONTENT OF TABLES AND FIGURES

Tables and figures can often contain the same kind of data. For example, Table 1 on the next page shows some of the data seen in Figure 1. If Table 1 had all the data from Figure 1, the table would be more than a page. More specifically, Figure 1 presents the atmospheric carbon dioxide concentration for each month, plus the annual mean, minimum, and minimum concentrations. So, the table of data for Figure 1 would be 15 columns across the top of the page (12 months, plus the annual mean, minimum, and maximum) and 47 rows down the page (actually 49 as there is one row for each year = 47 rows, plus a row for the heading of year and months in the columns, plus an empty row to separate the column headings from the data). So, a table of 15 columns x 47 rows of data has 705 data points, plus the space needed to separate rows and columns. One advantage of the use of Figure 1 is that all 705 data points are arranged in sequence, so the patterns of changing atmospheric carbon dioxide concentrations are clear with a quick glance.

Table 1. Atmospheric carbon dioxide concentrations (ppm) measured at Mauna Loa, Hawaii (Tans 2013).

Year	January	April	July	October	Annual Mean
1958		317		317	317
1968	323	325	324	320	323
1978	335	338	337	333	336
1988	359	354	352	349	351
1998	365	369	368	364	367
2002	372	375	374	370	373

Converting the data in Figure 1 into a Table requires some process or reducing the data. For example, Table 1includes data from every third month, at decade intervals. Although the Table 1 is roughly the same size as Figure 1, it contains only 27 data points, or roughly 5% of the data. In addition, some of the trends in changing CO_2 concentrations within a year and between years are not as evident in Table 1 as in Figure 1.

Authors must decide whether to use a figure or a table, and the decision will always rest on which form is more effective at communicating the message. For example, Figure 1 is effective at showing trends in increasing atmospheric carbon dioxide concentrations, and a table could never be as effective conveying such a trend. However, Table 1 is effective at showing how seasonal mean atmospheric carbon dioxide concentrations compare to annual mean concentrations, and does so more effectively than Figure 1. The atmospheric carbon dioxide concentration in October, since 1958, is lower than the annual mean concentration (Table 1).

Tables are also necessary for data organized in categories. For example, an author should use a table for listing specific locations of all the sites for measuring atmospheric concentrations of carbon dioxide, and the period of measurement (Table 2). As discussed for Figure 1, Table 1 is also a rare example of appropriate use of data taken from a website.

Table 1 demonstrates some of the issues that arise from taking data and citing a website. The source for Table 1 is the same as for Figure 1, the NOAA Earth System Research Laboratory. However, unlike Figure 1, Table 1 is difficult to find at the NOAA website. Fortunately, the NOAA website archives monthly and annual data for atmospheric CO_2 concentrations, so Table 1 can bereconstructed historic measurements.

Table 2. Names of stations with continuous measurements of atmospheric carbon dioxide concentrations, and their location, elevation, and measurement period (Keeling, C.D. and R.P. Whorf. 2004).

Station	Latitude (degrees)	Longitude (degrees)	Elevation (m)	Period Measured
Alert, Canada	82.0 N	62.0 W	210	05/85 – 12/01
Point Barrow, Alaska	71.3 N	157.3 W	11	01/74 – 12/01
La Jolla, California	32.9 N	117.2 W	10	11/72 – 10/75
Mauna Loa, Hawaii	19.5 N	155.6 W	3,397	03/58 – 12/01
Christmas Island, Kiribati	2 N	157.3 W	3	12/74 – 08/01
Baring Head, New Zealand	41.4 S	174.9 E	85	07/77 – 10/01
South Pole Station	90 S	0 W	2,810	06/60 – 10/63

Table 2 has interesting, important information in independent sets, and would make for a difficult figure. A consideration for a figure showing the same information might be a map of earth showing the locations of the monitoring sites, but providing the exact latitude and longitude, the elevation, and the sampling period would make for a map that is difficult to read.

8.4.6 FIGURE LEGENDS AND TABLE HEADINGS

Figures have legends that appear directly beneath, as shown for Figure 1. Figure legends begin with the number of the figure so it can be tied to a citation in the narrative. The figure legend also has text, with at least one complete sentence. Figures taken from another source should include a citation, to identify the source, as a parenthetical element in the figure legend. Authors should cite self-made figures in the narrative.

When using graphs, as in Figure 1, label both axes to identify the variables, time and atmospheric carbon dioxide concentration. Labels for axes on graphs should include units for each variable. When a graph has several lines, the writer must provide a way for the reader to understand each line. In some cases, lines can be different colors. In other cases, plot lines that connect different symbols, such as stars, boxes, dots, and circles. Whether lines differ by color or by symbols, provide a key to guide the reader. Present the key to figures either as an inset in the figure, or in the figure legend.

Tables have headings that appear above the table, as for Table 1 shown above. The table heading begins with the table number that links it to the citation of the table in the text. The heading of the table should explain the table content and the source of the information. Tables should not be split on two pages, but located close to the point of presentation, where it can be on shown on a single page.

CHAPTER 9

Top Ten Writing Tips

9.1   DO NOT PROCRASTINATE

Analytical writing is a rewarding adventure, often providing the foundation for important decisions. When committing to such a writing assignment, by far the best approach is to immediately begin the writing process. Delays in starting the writing process will only create problems.

Students learning analytical writing skills often struggle with time management issues as they cope with classes, jobs, social needs, and other commitments. Similarly, professionals with analytical writing responsibilities may have too many writing projects among many other responsibilities. Unfortunately, a common way to cope with competing demands is to delay work on writing assignments and responsibilities.

Analytical writing requires time to acquire information, develop objectives, draw conclusions, and produce a polished report. Delaying the start of a project ensures that the author will not have adequate time to succeed with the project. Worse still, starting a project with too little time to complete it is neither fun nor rewarding.

Writers who are part of a writing team must avoid the temptation to procrastinate, and be certain to set and meet target dates for specific writing products. When a single author within a writing team is late with their assignments, the outcome is that the report will almost certainly lack uniform formatting, consistent editing, and the desired level of integration.

9.2   NO "WHO DONE ITS?"

The best analytical writing reveals the big ideas, including the result of the analysis, early in the manuscript. Writers often err when they wait to reveal the big ideas, or the conclusion of the work, until the very end of the written manuscript. Such writing makes for suspenseful murder mysteries and thrillers, but those who read analytical writing often want to know the main thoughts and conclusions without searching through a thick manuscript. A good approach is to reveal the big ideas early in the writing project.

A good, analytical writer understands that professionals who make decisions about using resources must read many reports, journal articles, theses, business plans, and similar kinds of analytical documents. Such professionals typically do not read an entire document from beginning to end, and if the importance of the work does not appear until the end of the manuscript, the reader may never find it.

Readers often start with the introduction to see if the questions are interesting or relevant. The reader may then jump to the discussion section to assess the importance of the analysis, and then turn to the results section to focus on a specific figure or table to see if important data are robust. Finally, the reader may check the methods section to better understand the details of data collection and the logic behind the approaches. When the writer states key information, including the outcome of the analysis, early in the project, there is a greater chance of drawing a reader's attention to other parts of the document.

9.3   USE SIMPLE SENTENCE STRUCTURES

Analytical writing requires clear, simple sentences. Certainly when writing the first and second drafts, keep the sentence structure simple. With experience, writers can bring compound and complex sentences into the manuscript. However, relying on simple sentences, with easily recognized subjects and predicates will help readers understand the writing. Short sentences can result in a stilted writing style; even so, using simple sentences structures will help students avoid using long or run-on sentences that can be difficult or confusing. Examples of simple, clear sentences include:

- The data show air quality is improving.

- Improving air quality is reducing the incidence of asthma attacks.

- Improving air quality improves human health.

  An example of a long, difficult sentence is:

- The data show air quality is improving, leading to the reduction of asthmas attacks, and therefore improving air quality is also improving human health.

9.4   ASSOCIATE PRONOUNS WITH NOUNS

Inexperienced writers often fail to clearly associate pronouns, such as "it," "them," "they," and "those," with nouns. Because pronouns refer back to a specific noun, or replace a noun, the association must be clear and obvious. Avoid confusion by placing the pronoun immediately following a specific noun. Writers should not use pronouns at the beginning of a sentence or a paragraph because they are too far from the preceding noun. Some examples of poor pronoun usage include:

- The construction schedule for the building is late and makes it a problem. (The pronoun "It" could refer to the building or the schedule, and the meaning of the sentence is lost).

- They are planning to finance the new conference center. (Who are "They?").

- When you see the data in the graph, it becomes clear. (What does "it" refer too; the data or the graph?).

9.5   NO CONTRACTIONS, SLANG, IDIOMS, OR JARGON

Analytical writing is formal writing, and informal use of contractions, slang, idioms, and jargon is not appropriate. An important quality of analytical writing is that the thinking is clear and logical, and the use of language is precise. The use of informal language tools is appropriate in some forms of writing, but not in analytical writing. The use of informal language in writing is disrespectful to those in professional standing, and can confuse foreign readers.

9.5.1   CONTRACTIONS

Omit contractions, two words held together with an apostrophe, from formal, analytical writing. The following sentences contain contractions:

- "Differences between resource inputs and outputs aren't obvious."

- "A survey shows the new menu isn't going to attract Hispanic diners."

- "The commercial value of the watershed isn't equal to its value for ecological services."

In each of the three examples, the meaning of the sentence is understood. Yet, the informal shortening of two words into one contracted word has no place in formal, analytical writing.

9.5.2   SLANG

Slang is the language typically used in conversation, and includes non-standard expressions. For example, the sentence below contains two examples of slang that lack clarity for readers:

- **Example 1.** The flick *Gasland* documents public concern about fracking.

The word "flick" is slang for a movie, but is inappropriate for formal, analytical writing. Similarly, the word "fracking" is slang for "Hydraulic fracturing," and also should not be used in formal, analytical writing. The term "fracking" with time, may move from unacceptable slang into the formal vocabulary. However, the meaning of the word "fracking" is still more vague than the proper term "Hydraulic fracturing."

- **Example 2.** The decision to build the water treatment plant was dicey because the funding was not in place.

The slang word "dicey" has a vague meaning and detracts the reader's attention from the issue of the funding problem for the water treatment plant.

9.5.3 IDIOMS

Avoid using idioms, or words and expressions, which have meaning different from their literal content:

- **Example 1.** The new regulations defining air pollution standards are for the birds.

  The idiom "for the birds" is too vague to convey the author's meaning. The author may mean air pollution regulations are too lax, too stringent, too poorly defined, unattainable, unjustifiable, or have other problems. The author should be clear about the problems with the regulations, and to build ideas around specific issues.

  The sentence above may be a clever use of language. If the author is writing about air pollution regulations intended to protect birds, then the sentence takes on a specific meaning. The use of the expression, "For the birds." is especially clever if the author is being critical of the air pollution standards intended to protect birds.

- **Example 2.** If the new business plan fails, those responsible will pay the piper.

  The idiom "pay the piper" does not literally mean to pay the person playing musical pipes. Although the idiom has a meaning for the author, analytical writing should contain words that are clear and concise.

9.5.4 JARGON

Jargon takes many forms but is generally an attempt to take shortcuts with technical expressions. The key to coping with jargon is to define or explain terms that are understood only by experts. Although experts in a specific field understand their jargon, analytical writing will often have a wide range of readers and the text should be comprehensible to non-technical experts. For example, a feasibility study written by engineers must be compelling and appealing to financial investors. Specific examples of jargon include words used in science, medicine, and law that are rooted in Latin or other languages (isopleth), technical words created by combining words (e.g., Metabolomics), or words that are artificial constructs (e.g., blamestorming).

An important type of jargon is the acronym, or the use of synthetic words created from the letters of a commonly used string of words. For example, the National Aeronautics and Space Administration is commonly known by the acronym, NASA. One procedure for using acronyms is, at the first use of the word "string," use the full set of words, and then show the acronym as a parenthetical element:

- The investment in the National Aeronautics and Space Administration (NASA) will pay dividends.

After defining NASA at first use, using the acronym NASA throughout the rest of the writing is appropriate. Readers will understand that "NASA" stands for the National Aeronautics and

Space Administration, and not for other agencies and groups, such as the National Auto Sport Association. Although many people will recognize the acronym NASA, other acronyms such as ESA (Ecological Society of America or Endangered Species Act?) can be confusing and mislead readers.

9.6   AVOID PASSIVE VOICE

Sentences written in passive voice detract from analytical writing. Passive voice often results in sentences that are less clear and forceful than those in active voice. Sentences with passive voice have the subject as the recipient of action. Where possible, writers should revise these sentences.

Sentences in passive voice often have a tell-tale structure, making them easy to find.

- A form of the verb "to be" + a past participle

  ○ **Example 1.** The following sentence is in passive voice because the writer positioned the object of the sentence before the subject performing the action:

  "The manuscript was edited by Joe."

  "to be" = was

  the past participle = edited

Past participles often end in "ed" but can also have other endings, such as the words, "written, driven, and ridden."

The object of the sentence, "manuscript," comes before the subject that made the action, "Joe." Rewriting as active voice means simply putting the subject at the front of the sentence:

"Joe edited the manuscript."

  ○ **Example 2.** The following sentence is also in passive voice, but suffers from incomplete information.

  "The manuscript was edited."

As with example 1, above, the tell-tale sign is the "to be" verb, followed by the past participle, "edited." Here, though, the problem is simply that the subject of the sentence (the editors) are missing. The fix means identifying who did the editing and adding that information to the sentence, and then rewriting as follows.

"Joe and Jean edited the manuscript."

Rewriting sentences from passive to active voice helps readers by ensuring the sentence is simple and that information is in a logical sequence. Active voice also ensures a sentence contains all the necessary information.

9.7   USE SPELLING, GRAMMAR, AND EDITING TOOLS

Modern word processing software includes a wide range of tools that assist writers. Tools such as Spell Check, Grammar Checkers, and Track Changes are also excellent instructional aids. Obvious errors in spelling and grammar annoy readers, especially when software tools make it easy to correct such problems. Writers should budget time to use software tools that help ensure their manuscript is free from spelling and grammatical errors.

9.7.1   SPELL CHECK

Spell check is a tool that identifies typing errors and misspelled words. Writers should carefully observe words labeled by spell check, and take action. In some instances, a typing error is obvious, and easy to fix. In other cases, the writer may have a spelling error and can learn the correct spelling. The Spell Check tool will provide the correct spelling for several possible words, and the writer can quickly select the correct one.

Writers should not rely on spell check to find all spelling errors. Spell check may not help when typing errors and misspelled words form real words. In addition, writers can use the wrong word which is correctly spelled, such as "too" in place of "two." Spell check is helpful when used with a dictionary so writers are certain to select the correct word or term. Spell check can help teach both spelling and vocabulary.

9.7.2   GRAMMAR CHECK

Grammar check is a software tool that identifies problems with punctuation or poor sentence structure. When grammar check identifies a sentence, or part of a sentence, that is wrong or poorly crafted, writers can select alternate punctuation or sentence structures to correct the sentence. In general, grammar check will help writers learn to place punctuation, place the subject at the beginning of a sentence, avoid sentence fragments, and eliminate run-on sentences.

9.7.3   TRACK CHANGES

Track changes provides a way for writers to make revisions. Track changes allows an author to see the original text, along with edits from either the author or from a reviewer. When writers use track changes, they follow their own revisions. However, track changes is especially useful for writers who share their documents with a writing team, editors, and reviewers. Those reviewing or editing documents can alter the original document, while showing both the original and new versions of text. The author, upon receiving the revised manuscript, can either revert to the original version, or select the revised version. The author can also accept or reject specific changes. The track changes tool is a boon for those involved in team writing projects as the lead author can gather writing input from coauthors, annotating each revision in the text.

9.8   BREAKING WRITER'S BLOCK

Most writers, especially those who want to write well, spare no effort to prepare fully for their project. The preparation for writing often includes clearing the desk, eliminating distractions, and committing a block of time to do nothing but write. Unfortunately, after preparation, a writer's block victim then sits at a desk, fingers on the computer keyboard, and nothing happens. Time passes, and nothing happens. The author gets up, paces the room, gets a drink of water, and now refreshed, sits again at the keyboard. Nothing happens, and writer's block sets in.

The writer begins to get edgy. Time passes, and seconds turn to minutes, and more. Still nothing happens and the frustration begins to build. The writer took all the precautions necessary to ensure a productive writing session, and yet, no real writing is taking place. How does writer's block happen, how can the writer recover, and how can the project or assignment move forward?

Simply put, writer's block often results when the writer does not know what he or she wants to write or say. When an author knows what to write, the words quickly flow from the mind, through the fingers, onto the keyboard, and onto the computer screen with great speed. Authors in a productive mode can hardly write fast enough to keep up with the formation of new ideas and sentences.

The key for breaking through writer's block is to change from not knowing what to write to a condition in which words easily flow. The best way to cope with writer's block is for the author to design a process to figure out what to write, or say. Here are several steps to help overcome writer's block.

9.8.1   GET MORE INFORMATION

Sometimes writers do not know what to write because they do not have enough information to form a message. If so, the solution is to get more information. Sometimes more information is available online. Other times, needed information is a critical piece of data, a specific report, or journal article. If so, one solution is to go a library. Once a writer determines he or she needs more information, the only solution is to be aggressive and get it.

9.8.2   TALK TO PEOPLE

The writer may have lots of ideas and data, but still needs to assimilate information and concepts. A good approach is to talk to people about your ideas. A conversation with a specialist in the field is often useful. Similarly, a conversation with a friend, family member, spouse, sibling, parent, office mate, colleague, or fellow student can also be helpful. You can even talk with yourself to help clarify ideas about a topic. Often, breaking writer's block simply involves just talking about the project to verbally articulate what is important about the issue, to clarify the questions and the significance of the work. After authors say the words, it is easier to write the words.

If you are working on a team project, talk with others on the team. Writer's block may result if you are unsure of your specific writing assignment, and how it overlaps or dovetails with other sections of the project. Others on your writing team may have the same issues, and appreciate your taking the time to clarify writing responsibilities for everyone.

9.8.3   RETHINK THE OUTLINE

Writer's block can occur if the scope of the project is too large, or the analysis is too complex, and the author does not know where to begin. If the analytical writing project contains too much descriptive material to set the stage and provide rationale for the project, writers may not know where to begin or end. The answer for breaking writer's block for projects that are too large and undefined is to reconsider the outline and to create a project that has a solid starting point, middle body, and ending point. Although such simplification may shrink the project, the analysis often involves more than originally thought. If the completed project is too small, there is opportunity to add more development or new parts to the project.

9.9   WATCH THE UNITS

In general, there are two basic sets of units, English units and metric units. Once committed to one set or the other, be consistent. In some cases, it might be appropriate to use English units followed by the equivalent metric unit in parentheses.

Errors with units can create serious problems. A famous error in space exploration occurred from the misuse of units. In December 1998, NASA launched a $125 million unmanned satellite, the Mars Climate Orbiter. The satellite's objective was to survey the Martian climate and send the data back to NASA scientists and engineers. After traveling in space for more than 250 days, NASA flight controllers sent radio signals to slow the spacecraft and put it into a Martian orbit. All went as planned until the satellite disappeared. A NASA review board determined that navigational commands sent to the satellite were in English units, not metric units. The confusion in units caused the trajectory of the craft to miss the orbital path and put the satellite into a path to collide with Mars and fail the mission.

Although misuse of units rarely results in tragic loss, consistency is essential for clear writing. Analytical writing gives special consideration to the use of English units, metric units, and units of time.

9.9.1   ENGLISH UNITS

English units, also know as Imperial units, originated in England and their use spread throughout the Commonwealth. Those in the United States inherited English units and still use them today. Examples of some English units include:

- Length – inch, foot, yard, chain, mile

- Area – inches, feet, yards, acre

- Volume – jigger, cup, pint, quart, gallon, peck, bushel, barrel

- Weight– grain, dram, ounce, pound, ton

- Temperature – Fahrenheit (not really an English unit, but often used with them)

Most readers in the U.S. will understand manuscripts that use English units, so using them is convenient for American writers and readers.

9.9.2   METRIC UNITS

All writing in science and engineering should be done using the International System of Units (SI), standardized as the metric system. The SI system is the only universal set of units recognized and used around the world. Examples of SI units include:

- Length – meter

- Area – square meter

- Volume – cubic meter

- Mass – kilogram

- Temperature – Kelvin (degrees Celsius is not an SI unit but the SI convention endorses this unit)

The SI system of units uses prefixes to scale units and the range of prefixes used is expanding. For example, two decades ago, the common range of prefixes might range from milli (as millimeters or 10^{-3} m) to mega (as megagram or 10^6 g). Today, scientists commonly refer to nanotechnology where particles are 1–100 nanometers (10^{-9} m) in size. At the other extreme, climate change scientists who study greenhouse gas emissions commonly estimate parts of the global carbon cycle in petagrams (10^{15} g C) of carbon. Examples of prefixes include:

- nano = 10^{-9} (billionth), e.g., nanometer, nanogram

- micro = 10^{-6} (millionth), e.g., micrometer, microgram

- milli = 10^{-3} (thousandth), e.g., millimeter, milligram

- centi = 10^{-2} (hundredth), e.g., centimeter, centigram

- deci = 10^{-1} (tenth), e.g., decimeter, decigram

- deca = 10^{1} (ten), e.g., decameter, decagram

- hecto = 10^{2} (hundred), e.g., hectometer, hectogram

- kilo = 10^{3} (thousand), e.g., kilometer, kilogram

- mega = 10^{6} (million), e.g., megameter, megagram

- giga = 10^{9} (billion), e.g., gigameter, gigagram

- tera = 10^{12} (trillion), e.g., teragram

- peta – 10^{15} (quadrillion), e.g., petagram

9.9.3 UNITS OF TIME

Whether using English or SI units, units of time are always seconds, minutes, hours, days, weeks, months, and years. Writers can, but rarely use, prefixes with units of time. For example, a nanosecond is 10^{-9} seconds.

9.10 KEEP A NOTEBOOK OF WRITING ACTIVITIES

Writing in notebooks helps writers to gather and evaluate information before answering questions. Keeping a notebook is critically important in some situations. In general, a notebook should be a diary of daily activities. Proper notebooks help writers to see the rate of daily progress with projects. Notebooks are personal records for the author, but those who manage a project may also require some form of log or diary.

Notebooks for writing projects are the "gold standard" and the ultimate evidence for establishing dates for conceptual breakthroughs, invention, contributions, and the origins of invention. In short, the notebook annotates a project. If correctly done, notebooks are the ultimate evidence for settling disputes of who was first to an idea, provides the last word for legal cases to determine copyright, patents, or issues of plagiarism.

Many students and others who write analytical projects are not so concerned with establishing ownership of intellectual property, but should still develop skills at keeping a notebook. Notebooks are extremely valuable as backup for lost or destroyed materials and writing. Notebooks can also be a testing ground for exploring a new approach. Finally, a notebook is a way for an author to chronicle their progress in writing, thinking, and the evolution of the project, either alone or in a team context.

The notebook should include statements of purpose, bias, and motivation for the project, and include a project outline and timeline. Daily entries will help the writer to keep track of progress, and indicate which efforts are ahead or behind schedule. Examples of daily entries include a brief description of writing progress, names of computer files, efforts made to acquire information, ideas about how to evaluate facts or hypotheses, appointments for the project, relevant conversations and interviews, new ideas about the project, or decisions about solving problems. Notebooks should also contain results of analysis, data, and statements of conceptual development.

Writers should keep a separate notebook for each project. If a writer is participating in three projects, keep three notebooks. Important qualities of a notebook include:

1. bound with numbered pages;

2. start on page one, and do not skip pages;

3. dated entries;

4. entries made in ink, neatly;

5. no erasing entries;

6. entry errors shown with a single line through the error with comments about the error;

7. provide notes, sketches, hand-drawn graphs;

8. append raw data, computer printout, maps, etc.

A notebook becomes familiar and comfortable, like a pair of pajamas. Writers should look forward to using their notebook during the day, and be happy to make a personal entry.

CHAPTER 10

Ethics: Bias and Plagiarism

The idea of telling the truth seems simple. However, the concept of truth can be deceptive and difficult. The issues of increasing complexity include correctly interpreting real data, giving attribution, and coping with bias. At times, writers think they are being truthful, but are deceived by others, or by blinding bias. In other cases, authors are unethical when they fail to acknowledge others for ideas or when information presented is knowingly fraudulent.

The most severe outcome from the unethical misuse of information is to increase public distrust for science and scientists. With each episode of ethical breach comes further opportunities for public interest groups to discredit scientists and the emerging understanding of vital issues such as genetically engineered organisms, climate change, and nutrition.

10.1 CAREFULLY INTERPRET REAL DATA

Science is laden with many examples of scientific writing based on fabricated data. Using fake samples or fake data can be intentional or inadvertent, but in either case is an ethical breech that can ultimately lead to loss of an individual's credibility and reputation. Life long careers of hard careful work can be lost by use of fraudulent data. Those caught misusing data can lose jobs, grants, and patents and face imprisonment.

10.1.1 PILTDOWN MAN

The Piltdown Man hoax is an example of untruthful science that began nearly 100 years ago, and resolution is still not complete (Russell 2013). The hoax began in 1912 at an English archeological dig when an archeologist, Charles Dawson, found skull and jawbone fragments he thought were of human origin. Dawson took the evidence to Sir Arthur Smith Woodward, a renowned geologist and fossil expert of the day. Woodward and Dawson assembled the material collected from the excavation, and brought forward a skull and jaw they claimed to be the fossil remains of an early human. The two geologists named the specimen "Piltdown Man."

Piltdown Man was of great interest because the specimen appeared to be an evolutionary missing link between humans and apes. The skull included a cranium as large as modern man, with the jaw primitive in form; a perfect "missing link" between modern humans and apes. The Piltdown Man supported the idea that humans evolved from other life forms. Of course, evolution was a concept that first emerged from Charles Darwin in the 1870s, and scientists and the public were contemplating whether the theory of evolution might also explain human existence.

Although some scientists were immediately skeptical about the Piltdown Man find, the specimen was generally accepted as evidence for the evolution of humans from apes. The Piltdown Man hoax continued until 1953 when scientists used an array of techniques to establish that the skull was from the 1500s, and not an ancient human fossil. In addition, the jawbone was correctly identified to be from an orangutan, with teeth from a chimpanzee.

The Piltdown Man hoax had unfortunate impacts. Dawson died in 1916 and Woodward died in 1944, so the hoax outlived both founders. Still the credibility of both men was severely damaged, and even though Woodward made enormous contributions in archeology, he will always be known for perpetrating the misleading Piltdown Man specimen.

Most importantly, the fraudulent specimen led to confusion about the actual evolutionary pathways of primates. Confusion about the origins of the hoax persists, and will likely remain unresolved. One possibility is that Dawson and Woodward conspired and knew about the deception. Alternatively, the skull and jawbone samples could have been "planted" as a practical joke by others. If so, Dawson and Woodward should have been more rigorous about making a claim, as they both had the tools to recognize the fakery.

10.1.2   COLD FUSION

Stanley Pons and Martin Fleischmann are chemists who started a scientific uproar (Taub 1993). In spring 1989, they held a press conference at the University of Utah announcing the production of energy from a process termed "cold fusion." These scientists completed experiments and claimed they achieved nuclear fusion in glass vessels at room temperature and pressures (Fleischmann and Pons 1989). Many scientists regard cold fusion as impossible and dismissed the claims of cold fusion because the experiments of Pons and Fleischmann were not reproducible (Doe 1989). Regardless, the cold fusion experiments stimulated research investments and involved careers of many scientists. Although the furor around the concepts of cold fusion still persists, analytical writers can learn from the situation.

The process of nuclear fusion involves the combining of atomic nuclei accompanied by the release of energy. Nuclear fusion is important because it has the potential to supply global energy needs. Nuclear fusion takes place in the sun, and is the subject of expensive and extensive research at sites around the world. Until the announcement of cold fusion, research focused on using huge amounts of electrical energy to generate the heat, pressure, and magnetism necessary to induce and contain the reactions of nuclear fusion. In general, such efforts resulted in short fusion reactions yielding only a fraction of the energy necessary to make the reaction in the first place.

Several issues surround the cold fusion debacle and apply to the ethics and bias surrounding all analytical writing and thinking. The technical difficulties of high-energy fusion reactions make the cold fusion technology seem attractive; and perhaps too attractive. Pons and Fleischmann hastily organized a formal press conference to announce the importance of their cold fusion discovery before carefully evaluating and replicating their experiments. Had the two scientists been just a bit more rigorous, they would have avoided much of the ensuing controversy.

Pons and Fleischmann may also have had a bias about their work that removed the objectivity necessary for analytical writing and thinking. Such a bias can blind scientists who should seek fair answers to fair questions. Instead, bias can lead to experiments designed to prove a theory or technology. For example, the experimental designs behind the cold fusion experiments may have had conceptual flaws designed to prove the technology rather than to objectively test (or reject) the idea that the technology worked. Finally, research involving experiments, especially transformational experiments, should be published as a reviewed article so that others could comment on the work and replicate the experiments. In the absence of analytical writing to clearly spell out the background, rationale, objectives, approaches, results, and discussion that communicate the details about the work, the interpretation and replication of the original experiments simply is not possible.

Even though cold fusion is an example of a scientific hoax, there is no evidence that Pons and Fleischmann were intentionally misleading. Nonetheless, investors and research agencies spent millions of dollars, and scientists and engineers invested countless hours, pursuing the dream of cold fusion with little probability for success.

10.1.3 STEM CELL FRAUD

A South Korean scientist, Dr. Woo Suk Hwang, intentionally faked data and violated human ethics in one of the most important transgressions in the history of science. In 2004, Dr. Hwang published results from his laboratory showing he cloned human cells (Hwang et. al. 2005). The article (later retracted) claimed cloning human stem cell lines from specific patients. Dr. Hwang's claim of cloning human cells seemed credible because he had successfully cloned cells from other animals including pigs and dogs.

The news that Dr. Hwang cloned human cells was a significant advance because stem cells offer the promise of curing disease. Additional interest mounted because of the ethical issues of cloning human cells, collecting reproductive cells from humans, and the political issues that limit stem cell research in the United States. With all the controversy, the prospect that Dr. Hwang had the technology to clone human cells brought the scientist great accolades, and the country of South Korea enormous recognition for a profound, scientific accomplishment.

Problems for Dr. Hwang arose in 2005 when a collaborator, Dr. Gerald Schatten, at the University of Pittsburgh, broke off collaboration because of ethical concerns regarding the collection of human egg cells (Guterman 2006). Later, evidence showed Dr. Hwang paid women to donate eggs, a breach of ethics. Dr. Hwang also lied, claiming the women voluntarily donated their eggs, and then lied about his knowledge of paying the women. Problems amplified later in 2005 when further genetic analysis showed none of the claims of cloning human cells were reproducible, and that genetic data indicating successful cloning were fraudulent.

The lack of ethical science, including the process of collecting human ova and fabricating data, created an impact affecting many people, including Dr. Hwang (Normile 2009). Dr. Hwang and six of his research team lost their positions at Seoul National University. Dr. Hwang was indicted and convicted for fraud, embezzlement, and bioethical violations; he was not sent to jail. Perhaps more seriously, the scientific field of stem cell research still has great potential to deliver miraculous medical advances, but is still reeling from the impacts of Dr. Hwang's ethical breeches that will further delay this promising area of research.

10.2   GIVE ATTRIBUTION

Plagiarism results when authors do not tell the truth, and take claim of writing or the ideas of others. Academic institutions penalize students and faculty members caught plagiarizing from others, and the outcome can include failing a course, or expulsion from the institution. For those with jobs, plagiarism can result in the loss of reputation and termination from employment. Avoid plagiarism by using quotations and citations. By recognizing the contributions of others, authors gain credibility for their efforts and have nothing to lose.

Students are guilty of plagiarism if they use software to cut and paste text straight from other sources for a paper and showing the writing as their own. Students can also plagiarize if they buy or borrow term papers written by others. Finally, plagiarism results if writers take the substance of an idea from another without giving credit; simply changing a word or two does not absolve the writer from plagiarism.

Giving attribution is the difference between telling the truth and plagiarism. Giving attribution is an ethical way of acknowledging that authors do not create brilliant writing in a vacuum. Instead, writers build ideas, facts, and in the process acknowledge the work of others. Writers who cite others hope that their manuscripts will stimulate further new writing. In all cases attribution is an important way to document the progressive development of information and thinking, and in a sense is the ultimate notebook for a field.

One approach for giving attribution is to quote the writing or speaking of another. The quote is identified with quotation marks, and the attribution is completed when the source of the quotation is cited in the text. Quotations may be long, or short, but never longer than necessary. Quotations are not common in analytical writing, and typically are not more than one or two sentences.

Citations are another form of attribution. When the writer makes a statement of fact, the writer will give a citation that attributes the fact to a specific source. In so doing, the writer is sure that the reader can find the original source of the fact. Sources of ideas are also attributed to others with citations.

10.3 OBJECTIVITY AND BIAS

The best way for writers to cope with bias is to admit they have it. All people have bias, and admitting bias is essential to developing an objective perspective. Bias, or prejudice, is often associated with racial issues. However, bias can extend to virtually all facets of human interest. People develop bias about religion, gender, geographic locations, political parties, brand names, schools, and newspapers. An individual has bias that reflects their socio-economic background, and childhood experiences that formed values. By recognizing bias, writers can be more objective in analysis, because critical thinking and making wise decisions is best when founded in reality rather than misperceptions.

Writers, and those charged with making important decisions requiring analytical thinking, can be biased in many areas. For example, most people have strong feeling for subjects such as environmental issues, managing natural resources, economic development projects, land use laws and planning, planning restaurants, and financial investing.

When authors let their personal values guide thinking, one outcome discussed earlier can be "normative" science. In normative science, the author plants a preconceived policy outcome, secretly (perhaps even unknown to the author) in the design of the work. The result is that the author plays the role of a stealth advocate for a policy recommendation disguised as the outcome of the analysis.

Three scenarios follow in which bias is playing a role, but would be manageable by recognizing the bias.

1. **Environment and natural resources.** Jane is elected to City Council and has a voice and a vote in city planning. A developer wants to build a new housing complex at the edge of the urban growth boundary. Jane is a member of several conservation groups, is an avid bird watcher, and shops at an organic grocery store. Although Jane listens to the proposal for the housing complex, she is opposed to the development proposal even before the hearing. During the hearing, she does not really listen to the plans for development because she is thinking about writing an analysis of all the reasons for not issuing the development permit. The result is that whether the proposed housing complex is well conceived and has merit, or not, Jane will allow her bias to form her opinion, not the facts at hand.

2. **Financial investment.** John is a financial advisor and manages several large accounts. Portions of the accounts include stock investments. John drives a Mercedes, while strongly disliking cars made in the U. S. John thinks cars made in the U. S. are too big, are poorly made, and do not perform well. John never recommends purchase of stocks in a domestic car company and because of his bias, never explores if such companies make a good investment for his clients.

3. **Business decisions.** Fred and Sally have some money to invest in a restaurant opening in a new mall. Fred and Sally both know the mall developer, and those involved in filling the mall with businesses including the space for the restaurant. Fred and Sally invest in the restaurant, but then trouble threatens to unravel the deal. A Japanese chef wants to lease the restaurant, and has a television cooking show and successful restaurants around the world. However, Fred and Sally do not like Japanese sushi. Fred and Sally read the business plan for the Japanese restaurant, but turn their nose up at the sure moneymaker.

In each of the three scenarios, objectivity was put aside for the sake of personal bias. Admitting bias is an important first step in coping with personal preferences and values. Admitting and discussing bias is a critical exercise in both single-author projects and in group-writing projects. Writing teams from both the public and private sectors, or within academic institutions, can have bias centered on the identity of their corporation, business position, or field of science. Each contributor to the project should understand the bias that exists for individuals on the team. Once writers know where the bias exists, prospects for objectivity increase. If a writer decides they cannot be objective about a project, perhaps the writer should choose another project.

Connected to bias is understanding the motivation for doing the project. For example, a student's motivation might be that a high school teacher assigns a required writing assignment. If the student motivation is to please parents, impress the teacher, improve writing skills, and do well in college applications, the motivation to do well with the project is at a high level. Such motivation can fuel the effort for success, but can also become part of the bias or rationale the author brings to the project. On the other hand, if a student has no motivation to do well, they bring a different bias to the project that will reflect on the quality of their effort.

Motivation becomes a critical piece contributing to bias for professional writers. Some writers may respect their supervisors, and want to produce writing that contributes to success for all. Other writers may resent their supervisor affecting the motivation for the writing assignment. In either case, motivation is often connected to bias.

CHAPTER 11

Final Products

11.1 TERM PAPERS, THESES, REPORTS, AND PUBLICATIONS

Analytical writing produces the evidence of analytical thinking, and is a special product that stands over time. Analytical writing shows the development of questions, approaches, methods, analytical techniques, and the reasoning behind synthesis of information and its importance. The written statement endures over time, unlike seminars, lectures, interviews, or other non-written forms of communication. Writing is the ultimate product of analytical thinking.

Since analytical writing can take different forms serving the needs of an array of writers and audiences, writers must consider how to use finished products at the onset of the writing process. College students should know whether they are writing a term paper for a college course, or a thesis for the Honors College. Before beginning to write, graduate students should know whether they are writing a term paper for a course, a thesis, or a article for submission to an edited journal. Before deciding on the scope and the format for the endeavor, engineers, financiers, architects, and others who are writing as a part of their profession should understand whether the document they produce will be a report to the Board of Directors, or a book chapter.

11.1.1 TERM PAPERS

Many high school students, college students, and students in graduate school write term papers as part of the grade for a specific course or class. Often, the grade for a term paper constitutes an important, and perhaps the largest, element of a student's grade for a class. When writing term papers for classes and courses, students should fully understand the syllabus for the course, the role the term paper plays in the conceptual learning that is expected, and how the writing assignment contributes to the student's grade.

Instructors may require that term papers from single authors, in which case, the each writer owns a document. Alternatively, instructors may require that students work in teams. Team projects are valuable because they simulate the writing environment commonly found outside the classroom, allow for projects with scope larger than can be approached by single authors, and allow students to teach each other about facets of a single project.

Students involved in writing teams should work as a group with the instructor to ensure everyone understands the goals of the project and the guidelines for recognizing contributions from each author. Team projects, even when contributions from multiple authors are seamlessly integrated, should provide for an honest division of credit. Dividing credit can be difficult because some tasks may be either easier or more difficult than they appear. In addition, a difficult task for one team member may appear easy to another. Regardless, to prevent misunderstanding, communication among team members and the instructor throughout the entire project is essential.

Students should pay particular attention to the guidelines the instructor provides for the term paper. The instructions may specify acceptable topics or general themes for writing. The instructions may also require specific formatting for documents including margin size, font size, line spacing, format for graphical elements, formats for citations and references, and number of pages. Students at all levels should always discuss questions about term papers with the instructor.

11.1.2   THESES

A thesis or dissertation, here, is meant as writing done as part of an academic program for both undergraduate and graduate students. A thesis involves research, discovery, or the creation of a new intellectual product, and is typically more formal than other forms of analytical writing. An advisor and advising committee guide a thesis project that involves a concluding seminar, oral exam, and thesis defense. Authors submit the final version of a thesis to an institution that retains the document in a library for others to review.

The process of creating a thesis is a rewarding experience that allows the student to explain their new thinking on a subject, and to get credit for documenting the idea. The final thesis is the culmination of a considerable body of work, and distinguishes the author for the creative endeavor. The thesis can be a piece of work that allows the author to lay claim to concepts and intellectual products that define a life-long career path.

A thesis has only a single author. However, the student's advisor and the advising committee typically sign the cover page of a thesis following their review. In addition, the ideas in the thesis rarely come from the student alone and should reflect thinking, discussion, and other forms of input from the student's advisor, graduate committee members, and other experts in the field. So, even though the thesis will have a single author, authors of theses should acknowledge the important role others played.

Writing a thesis for an undergraduate Honor's degree, a Master's degree, or a Ph.D. degree can seem a daunting experience. Writing a thesis that will appear as a book or bound volume based on their own analysis and creativity intimidates many students. Yet, students who follow the ideas developed in this handbook will understand that writing begins with an outline of simple ideas. With thought and the process of gathering and evaluating information, the outline grows into a polished, final document. The best theses are those reviewed by outside readers, and go through an iterative process leading to progressive improvement. Theses written quickly at the end of a student's academic program can produce intense stress and lead to an unsatisfactory product.

A thesis can take many forms. In some cases the thesis is a single piece of analytical writing, that contains a single introduction, set of hypotheses, approach and methods, results, discussion, and reference section. In other cases, a thesis might consist of chapters that are set up to be articles the author submits to peer review journals. In either case, the academic institution involved has guidelines and formats for theses, and authors must be fully aware so they can comply. In addition to specifying format, writers must also be aware of deadlines for submission, be aware of the numbers of bound or unbound copies to produce, and options for submitting computer files rather than hard copy.

11.1.3 REPORTS

Professional scientists, engineers, and those in finance, business, or involved with resource management will find themselves writing reports. Reports can take many forms. Some reports will be for a small, specific audience and used to document progress, to sell a concept, or to provide a framework for decisions. Other reports will target a broad audience with distribution via the web.

Those writing reports in the public and private sectors should fully recognize the scope and the expectations of readers before gauging the effort and resources to invest in the project. In some cases, a report may be a three-page analysis of progress on a large project, with similar reports due on a monthly basis. Such reports should not evolve into massive efforts, but should nonetheless involve critical analysis of major issues including any extraordinary events, breakthroughs, or the prospects for future delays.

Working groups can retain reports not released for general reading. However, working groups can also produce large reports designed to impact a broad group of readers. For example, a single report, *Smoking and Health* (U.S. Surgeon General's Advisory Committee 1964), increased public awareness of the health risks from smoking leading to new efforts to help the public better manage risks from smoking cigarettes.

Another report from the Intergovernmental Panel on Climate Change, *Climate Change 2007: Impacts, Adaptation, and Vulnerability* (Parry et al. 2007), established clear trends in global climate attributable to human activity. The scope and depth of the report set the stage for discussions of mitigating and adapting to climate change in all nations. The profound nature of the report led the IPCC to share the 2007 Nobel Peace Prize with Mr. Al Gore.

Both the reports on smoking and on climate change are large, valuable examples of analytical writing that involve writing teams, and years of effort. The reports each considered thousands of research papers, and produced a synthesis that transformed policies, economics, and behaviors for people around the world.

11.1.4 PUBLICATIONS (JOURNAL ARTICLES, BOOK CHAPTERS, AND BOOKS)

A wide range of magazines and journals, book chapters, and books publish analytical writing. Each publishing outlet has unique features.

Magazines and journals may take unsolicited manuscripts and consider them for publication. Commercial magazines have writing and editorial staffs, and may not accept unsolicited articles. Commercial magazines may buy articles from freelance writers, or negotiate contracts for authors to produce articles. Most freelance writers negotiate the details of the writing in advance of committing to the work.

Professional societies and publishing companies publish journals for readers in specific areas of science, engineering, medicine, environment, business and economics, and in virtually all areas involving scholarship. Many professional journals have an editor, an editorial board, and review panels. Publishers of such journals invite authors to submit manuscripts for review and publication. Upon receiving an article for review, the editor first determines if the subject of the manuscript is appropriate for the journal. If so, the editor assigns the manuscript to a member of the editorial board who will supervise its review.

The managing editor will send the manuscript to a number of experts in the field who will read the article, and provide opinions explaining whether or not to publish the article. The outside readers will justify their opinions by providing the managing editor with detailed comments explaining the strengths and weaknesses of the article.

The outside readers are peers of the author, have expertise in the author's academic field, and are referred to as "peer reviewers." In most reviews, the outside peer reviewers remain unknown, or anonymous. Maintaining anonymity allows the reviewer the freedom to give an opinion without risks or fears of retribution from the author. In an ideal world, the fact that the reviewer is unknown means they have nothing to gain or lose from publishing the article, and the reviewer's opinion therefore is strictly a comment about the value of the thinking, the analysis, and the writing. The anonymous peer review is the highest standard for assessing the quality of a manuscript, and publishing in peer reviewed journals is the gold standard for quality writing.

Upon receiving the written reviews and opinions from expert readers, the managing editor will make a decision about the manuscript. The editor may decide to accept the submission for publication, to accept a manuscript after revision, to redo the review of the manuscript following revision, or to simply reject the article.

Once a manuscript is accepted for publication, much work still remains. The publisher must use the manuscript submitted by the author to produce galley proofs of the article. The galley proofs show the layout and appearance of the article as it will appear in the journal. The author must carefully review the galley proofs to find errors, or whether problems exist with the layout of the figures, tables, and pieces of text. Also, if the article contains a table, figure, or some other feature taken directly from a source that is protected by a previous copyright, the author must request and receive permission from the copyright holder to use the material. The author will usually sign forms giving the publisher the copyright for the article. Finally, the author is typically not paid for the article, and may have to pay for reprints.

Edited books have qualities of both journals and books. A single editor, or an editorial board, may organize the volume. The editing body will create a specific series of chapters for the book, work to find authors for each chapter, and provide guidance to writers to be sure the chapters create an integrated volume rather than a series of loose, unrelated chapters. The editors will also establish writing deadlines, reserve the right to edit the writing in each chapter, and make final decisions on the format for narrative and graphical elements. The editors may choose to have one, long reference list at the end of the volume, or to have reference lists at the end of each chapter.

Authors receive credit for their chapter, and the editor receives credit for editing the volume, but not for writing a book. The editing body may choose to write an introduction, or a chapter, and if so, gets appropriate authorship credit. Some edited books are produced only once. However, edited volumes can also be a standing series, and similar to a journal with the hard binding of a book. The writing for a chapter in an edited book is typically of a more synthetic, general nature than a traditional manuscript submitted to a refereed journal. Some edited books include the process of sending each chapter out for anonymous peer review.

11.2   SEMINARS, SPEECHES, LECTURES, ORAL REPORTS, AND THESES DEFENSES

The opportunity to speak to a group is a special moment, and to be taken as a privilege. The members of the audience make a special effort to be present, must travel to the forum, and may pay a fee to attend the presentation. People will only go through the effort, and endure the expense, to attend an oral presentation if the speaker brings a message with value to the audience. Those engaged in analytical writing and strategic thinking are producing intellectual products with value, and should plan on giving oral presentations.

Oral presentations have the advantage of giving a group of people a single message at one moment in time. However, once said, the words are gone. People rarely visit archives of taped or filmed oral presentations. Consequently, documenting the text in a lecture, seminar, or oral report is difficult. Oral presentations never substitute for a written document that contains largely similar material.

Delivering an oral presentation gives the speaker a chance to develop a personal relationship with the members of the audience. Even if the presenter does not know a single person in the audience, a relationship will form. Speakers should reveal their passion for the topic, and aspects of their personality. An important reason people attend oral presentations is to get a personal glimpse of the speaker. When making oral presentations, be mindful of the impression you want to make with your appearance, clothing, and the way you carry yourself. The audience came to both see you and hear your message.

Those who consistently make effective oral presentations do so only through experience. Although some speakers are naturally more gifted at oration than others, everyone improves with practice. Courses in public speaking are not necessary to become accomplished at oral presentations. However, such courses can be a great aid and facilitate the development of speaking skills.

11.2.1 COPING WITH NERVES

All speakers get nervous, especially those making their first presentations to an audience. Getting nervous is a perfect reaction to a new, stressful situation. Writers should never let the fear of public speaking result in refusing assignments, failing part of a course, or creating any disadvantage. Authors should look for opportunities to explain their work to an audience. If necessary, counseling can provide additional resources to overcoming the fears of speaking in front of an audience.

One approach is to use nervousness as the source of energy to prepare an oral presentation. Any student or executive who will give an oral presentation or speech should prepare as any professional performer. Professional performers are comfortable with an audience because they know exactly what will happen on stage. Singers know their songs well, and will practice to be sure of the melodies and words. Musicians rehearse their scores so they know their parts of the concert. Similarly, those engaged in analytical writing and thinking can prepare in advance so they know exactly what will happen when they go on stage or to the podium. Writers can channel nervous energy into the work necessary to prepare and rehearse the presentation so that the outcome of what happens during the seminar or speech is no longer in question.

Outlining and rehearsing an oral presentation is the best way to ensure speakers know what will happen when giving a presentation. By outlining, the speaker will know the organization and the content the presentation. By rehearsing, the speaker will know the length of the presentation and the sequence and layout of graphical aids. Outlining and rehearsing changes the message from the written, analytical document, to a message so familiar that the presentation becomes as comfortable as favorite shoes.

One approach is to write out the text for a clear, concise introduction. If using slides or graphical images, write down the two or three most important ideas about each. Then, write out the text for a clear, concise summary or ending statement. After several rehearsals, the speaker should be able to easily deliver the introduction, run through the slides or images, and deliver the summary without referring to any written text or notes. Never read an oral presentation to an audience.

Speakers often use PowerPoint images for their seminars and other oral presentations. PowerPoint images can be effective communication aids. However, the overuse of such images, and presenting images with too much narrative, can result in reading the material to the audience. The words from the speaker are the message, not the images on the screen. More specifically, those in the audience who cannot see, such as those who are blind or cannot see the screen, should fully understand the presentation because the visual material is only a communication aid.

Professional performers also become familiar with the stage to eliminate uncertainty about what will happen in front of the audience. If a dress rehearsal is not possible, speakers should arrive well in advance of the presentation to learn about the room, the stage, and the podium. Locate the slide projector, or computer needed for the presentation, learn how the equipment works, and how to change images on the screen. Learn how the lights in the room work so you can control them to your advantage. If there is a microphone, learn how to use it before the presentation. If there is no microphone, adjust the volume of your voice and pace of delivery to be sure the entire audience can hear you. Learn about the position of any screen you will use in relation to any podium, and whether there is a pointer or not. Notice if there is a clock you can see from the podium so that you can keep track of time without having to glance at your watch during the presentation. Within a few minutes, a speaker can become familiar and comfortable with the setting for the presentation.

11.2.2 THE AUDIENCE AND THE TIME

Effective public speakers understand their audience. For example, a member of the School Board may prepare the analysis of a plan to convert the public school system from a traditional nine-month school year to a full calendar school year. The oral presentation to the School Board and School District Superintendent would likely be different from the presentation of the same material to concerned parents and students. In another example, an ecologist completing an analysis of the biogeochemistry of a coastal wetland would make one type of oral presentation to the Wetlands Ecological Society, and a different presentation to the commercial fisherman in the study area.

Although understanding those who will be in the audience seems easy and straightforward, predicting the size, level of interest, and expertise with the subject matter may be difficult. Speakers should make no assumptions about the audience; before preparing the oral presentation, make the effort to know who will be listening.

Speakers must know their allotted time. Seminars typically last 50 min. Student may have 15 min or less if several presentations are to fit in a single class period. Oral presentations to advisory boards may be 30 min. In general, prepare an oral presentation that uses about 80% of the allotted time so the audience can ask questions at the end of the presentation. Avoid running longer than the allotted time by watching a clock during your presentation. If running out of time, omit narrative from your presentation, but never omit the summary. Avoid the temptation to fit a time slot by simply talking faster, but instead plan your message to fit the available time.

11.2.3 ORGANIZING ORAL PRESENTATIONS

Organize effective oral presentations in the same outline form used in analytical writing. Well-organized oral presentations will have the following elements.

- **Introduction.** Present the title, set the stage, explain the need and importance of the subject, use background information to distill specific hypotheses or goals.

- **Approach and Methods.** Explain the approaches used to resolve hypotheses or to achieve goals, and outline the methods used.

- **Results.** Present the outcome of the work within the context of the hypotheses or goals.

- **Discussion.** Highlight the significance of the results. Do the results resolve the hypotheses, or are the results equivocal? What more needs to be done? What are the next pivotal questions?

- **Summary.** Briefly review the importance of the hypotheses and goals, the approach and methods, the results, and the significance of the results.

- **Acknowledgments.** Close by thanking those who helped with the work.

11.2.4 FIELDING QUESTIONS

A key component of any oral presentation comes at the end, when the members of the audience can ask questions. Many speakers fail to properly prepare for questions, yet this is the time when if unprepared, speakers can have problems. A well-presented oral presentation will fall short if the answers to questions are not sharp. Speakers will struggle with questions if they put all their energy into prepared remarks, are tired at the end, and questions catch them at low energy and surprised. On the other hand, sharp answers to questions can truly distinguish a speaker and the message.

Prepare for questions from the audience long before delivering an oral presentation. Upon deciding the material for a presentation, ask "What questions will come from the audience?" Write down the obvious questions that come to mind. Discuss the material with a peer prior to the presentation, and get their ideas about what questions to expect. Once you have a feeling for several questions, you can prepare the answers in advance. Also, reshape the content of the oral presentation to either invite or avoid certain questions, with both strategies being to the speaker's advantage.

In some cases, a member of the audience will interrupt your presentation with a question, presenting the speaker with a dilemma. When a speaker answers the question, it may be an invitation for more questions during the presentation. If questions come up repeatedly, the presentation can lose continuity. Alternatively, the speaker can quickly answer the first question, and then suggest other questions are welcome at the end of the presentation.

11.2.5 TEAM PRESENTATIONS

A team of individuals engaged in an analytical writing project can give an oral presentation to report on progress or to deliver an overview of the project with conclusions. Deciding how to cope with team presentations is difficult, and requires discussion among all team members.

The most efficient way to deliver a message, whether for a team project or for a project done by one individual, is to have a single spokesperson. One person delivering the message gives continuity and synthesis for the message that can be lost when several team members share the presentation. Before deciding on a single representative to deliver the message developed by a team, discussion issues include:

- **Choosing the speaker.** If one person presents the work of the team, the team must decide who will represent them. The natural tendency is to pick the person who formally or informally led the team. However, a better choice might be the team member with polished speaking skills, or an individual with the time available to adequately prepare the presentation.

- **Presentation by a team.** A team may elect to have all team members share the stage when presenting a project. In some cases, instructors may require each team member to participate in an oral presentation. When team members share the stage, planning is necessary to ensure a cohesive presentation.

A team presentation is easy to divide if each speaker did a specific piece of work. A presentation divided by individuals, each with a different theme, needs an introduction explaining the overarching goals and the connections between the pieces of work. A summary refocusing on the synthesis of the work is also important. Each individual involved in the presentation should be available for questions from the audience.

Regardless of the approach, a team must discuss and agree about how to produce the presentation. If there is no agreement, or if a decision cannot be reached, the team should consult the instructor for the course or the administrator supervising the team project. In general, teams are better off when they agree about the approach for presentations, rather than seeking input from outside sources.

Once a team decides on an approach for the presentation, there must be agreement on the content. Each team member should write an outline showing the planned content so everyone understand the entire presentation. The team members should meet to review the outline, and have an opportunity to make suggestions. All team members should participate in a full rehearsal to see the presentation, and to ensure the messages are coherent and not contradictory. If a single individual, or subset of the team, is making the presentation, similar team review is necessary and the speaker must take care to fully identify team members and their roles.

11.2.6 ORAL EXAMINATIONS

Oral Examinations can be a part of course assignments for students or a part of a thesis defense for both undergraduate and graduate students. Whether the oral examination is part of a course, or part of a thesis defense, students will feel great pressure.

Students taking an oral examination for a course assignment may receive either a letter grade or a pass/fail assessment. Instructors must make grading options clear for students, and fully explain the scope and content of the oral examination.

Pass or fail grades are most common for a thesis defense; in some cases, students can retake a failed examination, and in other cases, not. For a thesis defense, years of research culminate with the thesis, and the oral examination. In general, students do not get to the thesis defense stage of a graduate program if there is strong doubt about a strong thesis defense. So, even though the thesis defense can be stressful, students receiving good advising and mentoring should be confident of success.

Oral examinations generally begin with the student presenting a seminar summarizing the content of their thesis or class project. The oral seminar is followed by general questions from the audience, and a subsequent discussion with an evaluator or evaluation committee. The oral presentation is usually 50 min for a thesis, or less for a class project. Often, graduate students assemble a panel of peers to create a mock oral examination over the thesis in order to anticipate questions and to practice answering them.

Students should fully prepare for both their seminars and oral examinations covering their class projects or theses. The structure of the seminar should reflect the content and structure of the project report or thesis. Students should practice their seminar in front of peers and other advisors. The key is to produce an analytical report or thesis, and then to create an equally analytical seminar that is parallel in structure and content.

CHAPTER 12

Evaluating Analytical Writing

A critical, acquired skill for all readers is to evaluate writing. Increasing analytical writing skills leads to increasing reading skills. In short, the more papers and reports a person reads, the better they become at evaluating the quality of thinking in the manuscript. For example, experienced writers read to find and adopt approaches and techniques used by other writers. Such attention to the writing of others improves skills in evaluation and advances writing and thinking skills.

12.1   EVALUATION OF WRITING IN THE WORK PLACE

The process of evaluating analytical writing is a constant feature for making decisions in the work-place. Evaluation of analytical writing plays a key role in any business transaction, environmental assessment, project proposal, or any investment of time and resources that requires rational thinking.

Writers should remember that experienced readers of analytical writing rarely have time to read a document from the title page to the end matter. Rather, readers will often look at the title and then the summary to be sure the content of the article is relevant. The reader might then go to the discussion section of the article to see the importance or significance of the findings. If the findings are relevant and significant, the reader might then go to the results section to see whether there is compelling evidence for conclusions. If so, the reader might then go to the methods section to learn more about the choice of techniques and methods.

The reader might put down the first article, and then read several more articles while searching for areas of both agreement and disagreement. Easy decisions result when all evidence from analytical writing points in a single direction. When analytical writing points in different directions, those making decisions must invest effort to understand the basis for differences before deciding how to proceed. Most who make decisions seek writing that shows important differences in predictions or outcomes. Those employees who can excel at making complex decisions have special value, especially when they can clearly explain the reasoning that leads to positive direction and outcome.

Although the evaluation of analytical writing has no single criteria for defining success or failure, those experienced in analytical writing look for several features that signify a well-crafted article: concise writing with compelling rationale and objectives, author bias, gaps in reasoning, connecting goals to evidence and conclusions, effective graphical elements, and careful use of reference materials.

12.1.1   LOOK FOR CONCISE, COMPELLING WRITING

Experienced readers look for concise writing that is fast, efficient reading. Characteristics of concise writing include simple, declarative sentences, well-crafted paragraphs with topic sentences, and tight organization. Concise writing expresses ideas only once, and in the correct location within the document. Concise writing avoids slang, jargon, or vocabulary or expressions that may be ambiguous or vague. Concise writing distills goals from a compelling background statement.

12.1.2   WATCH FOR BIAS FROM AN AUTHOR

Experienced readers keep in mind how an author's bias might affect the analysis. For example, if a banker is reading a business plan from an entrepreneur requesting a loan, the bias from the author is to sell the plan to the banker, and the bias is obvious. Even though authors all have bias, the author may still have an objective analysis, so the business plan submitted to the banker may represent a solid venture. Readers should always be aware of author bias that results when the writer stands to gain resources, stature, or aims to influence an issue.

12.1.3   WATCH FOR GAPS IN REASONING

Gaps in reasoning invite critical readers to discount or dismiss an article even if there are meritorious elements. Experienced readers look for gaps in reasoning that can occur at several levels. For example, the specific goals or hypotheses should logically fall from the background information. If not, there is a gap in reasoning. In addition, the approaches, results, and discussion should pertain directly to the rationale and goals stated in the introduction. If the introduction identifies three goals, but the writing is unrelated, gaps in reasoning appear. Experienced readers look at all the steps in reasoning within a document.

12.1.4   CONSIDER THE REFERENCE LIST

Readers experienced in evaluating analytical writing may spend considerable time with the reference list. Weak reference lists might cite only a few, general references such as books. If so, the reference may not be to the original source, general in nature, and therefore, weak.

Another reference list red flag rises when too many citations come from a single author. For example, if there are twenty articles in the reference list, and fifteen of the references are from a single author, the reader will wonder whether the analysis is really considering all the issues.

Readers may want to know the author's experience in the field. If appropriate, authors should cite and refer to themselves in the text, and use their relevant articles in the manuscript and reference list to show some authority with the topic. However, authors should not overload the manuscript with excessive citations and references for their own work. Such overuse of citations raises questions of balance and objectivity.

Avoid citing and referring to websites. If sources in the reference list are web sites, the experienced reader will question the author's depth of understanding, and the rigor of the analysis.

12.2   FORMAL EVALUATIONS AND GRADING

Instructors should carefully evaluate analytical writing submitted as either part or all of a graded course. The evaluation of analytical writing by students should be part of the teaching mission for the instructor, and all activities associated with the writing assignment should provide educational value. Evaluating student writing should be one of the most important aspects of teaching and learning.

Instructors must fully explain the grading criteria and process to students, and present the explanation for evaluation and grading when announcing the assignment. Students have a legitimate complaint if they do not understand the grading system before beginning the assignment, after submitting the assigned project, or when receiving evaluation of the work.

Some assignment evaluations may be non-graded, graded as pass/no pass, or contribute to the final course grade. Regardless, instructors should evaluate analytical writing with an objective process that clearly explains both the evaluation criteria and process, and advances student understanding of the strengths and weaknesses of their work.

12.2.1   EVALUATION CRITERIA

Students and instructors are typically comfortable when a point system is used to evaluate analytical writing. Teachers and instructors should settle on a scheme, and fully explain it to students.

A strong approach is to evaluate the writing on the basis of four criteria, each of equal importance. For example, the instructor might make a writing assignment, and announce grading papers by evaluating categories including: organization, content, format, and graphics. If a manuscript is strong in all four equally important categories, it is a meritorious project. As students attend to each category, they cannot avoid creating a quality, analytical article. The four criteria are useful for evaluating all assignments, including iterative drafts, team reports, and seminars.

Organization (25%)

Content (25%)

Format (25%)

Graphics (25%)

Total (100%)

Teachers can design single evaluation sheets that shows the name of the instructor, title of the course, and provide blank spaces provided for entering the student's name, submission date, potential total points, and the specific assignment reviewed. The evaluation page can also list the four evaluation criteria spaced so that the instructor can write comments for each evaluation criteria. In addition, show the total points received in a space on the evaluation form. Of course, the most important teaching and learning depends on the nature of the comments teachers write in the evaluation form. Constructive comments are always more useful than negative criticism.

Instructors can help students by explaining some of the elements in the grading criteria so students will understand their scores. By doing so, the instructor is actually teaching students about analytical writing, and the expected level of thinking and writing. For example:

Organization (25%)

Does the Introduction set the stage with rationale leading to compelling objectives?

Is the use of headings and subheadings appropriate?

Do the ideas develop in a logical way?

Are there gaps in reasoning?

Are the results clearly presented?

Is the importance of the results assessed?

Does the paper have all necessary parts?

Content (25%)

Do concepts emerge?

Is there any information?

Is the information presented appropriate for analysis?

Does the author work with information to create new ideas and concepts?

Are ideas and other material properly attributed to others?

Are results properly interpreted and aligned with goals?

Does the title page contain all the appropriate information?

Format (25%)

Is the narrative clean of errors in spelling, grammar, and syntax?

Are margins uniform?

Is there a consistent citation style?

Is there a consistent reference style?

Are all references cited and all citations referenced?

Is there a consistent font for the text?

Is there proper use of headings and subheadings?

Are pages numbered?

Graphics (25%)

Are graphics presented with proper citation in text?

Are graphics properly presented in the narrative?

Do graphics have proper legends or headings?

Are graphics well crafted and easily understood?

Is attribution given for graphics taken from other sources?

Did the author create any graphics?

12.2.2   EVALUATION PROCESS

The process of evaluating analytical writing can be simple. For example, the students may submit a final document for a single grade. Alternatively, the evaluation process can be complex, with iterative submissions of manuscripts showing progressive improvements.

Simple Evaluation Process. A simple evaluation process, in which each student submits a final draft by a deadline date and time, can place a large, stressful burden on students. The burden is especially large if the writing assignment is an important part of the course grade. When students have an analytical writing assignment, and only submit a final copy, they should find ways to informally evaluate their earlier drafts. Student writers can seek reviews from other students or from experts outside the class environment.

Students in the class can informally assist each other by evaluating early drafts of manuscripts using the same grading criteria used for the final version of the paper. Above all, students must avoid working alone on a manuscript, and then submitting the first draft as a final version. Analytical writing always improves when authors can rewrite in response to suggestions from critical readers. Writers should welcome comments from critical readers, even if there are suggestions that authors should make large changes, such as backing away from ideas, either broadening or limiting scope, or working more diligently on the organization, writing mechanics, or graphical elements.

Iterative Submission Process. An iterative writing and evaluation process is fulfilling for both students and instructors, and is the process of analytical writing outside the classroom. The first evaluation product might be the outline. Subsequent evaluations could include a first draft of the manuscript, a second draft, and a final draft. Except for the outline, the grading criteria (organization, format, content, graphics) can remain constant for each iteration of the manuscript.

The iterative evaluation process allows the students and the instructor to see the progressive development of an idea and a manuscript. When an instructor can see the progressive development of a manuscript, there is less likelihood that students copied or plagiarized the document. Also, there is less possibility that the student is submitting a manuscript originally produced for another course. Finally, reviewing early drafts of writing assignments encourages students to begin work throughout the assignment period, and discourages procrastination. Early drafts of the manuscript will indicate which students are progressing with the assignment, which students need assistance in specific areas, and which students are not writing at the appropriate level.

Students can trade papers to evaluate the first draft of manuscripts for each other. The evaluation criteria for the first draft are the same as for the final draft allowing the students to learn more about the evaluation criteria by applying it to the writing of others. Consider using a low point total value for the first draft, especially when student graded.

Before beginning the second draft, authors should have adequate time to consider and act on comments following evaluation of the first draft. The instructor should evaluate the second draft and provide comments and points for each of the grading criteria. The total points for the second draft can be greater than for the first draft, but less than for the final draft. For example, if the first draft evaluated by students is worth 10 points, the second draft evaluated by the instructor might be worth 40 points.

The instructor must be sure to return the evaluated second draft of the manuscript to the student with enough time to make revisions for the third and final draft, and for the team report. Students should also have time to make further revisions of their own invention. To facilitate evaluating the second draft, the instructor should consider the grading criteria, and avoid making detailed, line-by-line corrections of grammar throughout the entire manuscript. Instructors can carefully edit a paragraph or two to help students see the standard of writing, including spelling and syntax, expected throughout the entire manuscript.

Final drafts submitted at the end of the academic period or year are typically worth more points than the sum of the points for the first and second drafts. The instructor should evaluate each student's manuscript with care and attention, fill out an evaluation sheet for each manuscript, and make the evaluation sheet and manuscript available for each author.

Evaluating the Team Report. One approach to building a team report in a high school or college class is for each student to contribute a chapter, so that students no longer fear that some will do more work than others. When each student contributes a chapter to a report, the work from each student is obvious and each student is fully accountable.

Students submitting a team report should also write a comprehensive introduction and conclusion that embrace the scope of the contributed chapters. The evaluation criteria for the team report can be the same as for chapter drafts. However, the point value for a team report can be low, as some students may not produce quality chapters, or may drop the course leaving the team report with missing elements. Still, all students finishing the project will take great pride in assembling a team report, and presenting a substantial document.

Evaluating Oral Presentations. The criteria evaluating manuscripts are also appropriate for evaluating oral presentations. If the instructor uses the equally weighted criteria of organization, content, format, and graphics for writing, students will quickly understand how these criteria translate and apply to an oral presentation. For example:

Organization (Oral presentations have the same elements considered for writing)

Content (Oral presentations have the same elements considered for writing)

Format (Oral presentations have elements different than considered for writing, for example:)

> Is pace of delivery appropriate?
>
> Is voice volume appropriate?
>
> Is English usage appropriate?
>
> Does student have appropriate posture and appearance?
>
> Does student comply with time limit?
>
> Does student deliver presentation free from reading notes?

Graphical Elements (Oral presentations have elements different than considered for writing, for example:)

> Are the figures and graphs easy to see on the screen?
>
> Are the figures and graphs properly constructed?
>
> Is the oral explanation of the information on tables and graphs accurate?
>
> Is the oral explanation of the information clear and comprehensible?

CHAPTER 13

Classroom Excercises for Teachers and Students

13.1   IDENTIFY INDIVIDUAL AND TEAM TOPICS

The instructor should spend much of the first day of class explaining the goals of analytical writing and thinking to students. When students first understand the goals of writing in a course, that is the ideal time to allow them to choose their topics. Letting students choose their topics is an excellent way to let their passion for an issue be the fuel that drives them through the course. When students get to choose their topic, they will do more than expected, and will own the project.

A course with virtually any theme can generate enthusiasm by letting students actively participate in forming teams and selecting topics. For example, if the course is a university capstone course in environmental science, the instructor can simply ask students what are the environmental topics of the day. Students will invariably come up with topics, such as: air pollution, drought in the Midwestern U.S., extinction rates, etc.

The instructor can write the topics on the board as students rattle them off. After 10–15 min, students slow in their additions to the list, and eventually no new topics come forward. Let the students look at the board for a minute or two, and ask each of them to mentally find three topics that that would make them happy.

After students make their mental choices, the instructor can pick any topic on the board, and ask for volunteers to form a team. The instructor might choose "air pollution" to see if there are 4–7 students who want to work on the air pollution team. Those students should immediately move to a place where they can share contact information, and begin discussing when they will next meet, and ideas for the team project. If 4–7 students raise their hands, they become a team that will develop a team project on air pollution. If 0–3 students raise their hand, there is not enough interest from the class, so the instructor can erase the topic from the board. If more than 7 students raise their hand, consider forming two teams on air pollution, chances are the two teams will go in quite different directions.

The instructor can then move on to the next topic on the board, such as "drought in the Midwestern U.S." Again, ask for a show of hands from students who would like to be on the Midwest drought team. Form a second team if enough students choose that theme, or erase the theme from the board if student interest is too low.

Continue the process until all the students are on a team, or until a subset of students do not have a team and all the topics are erased. Get topics from remaining students, and repeat the process. At the end of the period students have an introduction to the course, a team, and ownership of a topic or theme so they can begin collecting information and reading.

13.2 PLAY ROLES

Instructors in high school and in higher education can play multiple roles. Of course, instructors are responsible for the course content, management, and grading. Students are responsible for assignments and achieving course goals. But instructors and students can also play other roles.

Instructors can play the role of project manager for a consulting company. The students also take on the role of consulting company employees. The project manager needs recommendations from employees for new projects, business applications, and reports that open to new opportunities and innovation. Without such reports, the consulting company cannot succeed.

The role playing should reflect the interests of the students and the instructor, and for some teams, the consulting company might be centered on environmental monitoring. In other cases, the instructor might take on the role of the head of the U.S. Department of Energy, a position in the National Institutes of Health, or the World Wildlife Fund. Regardless of the business or agency, the goal is to get the students to think as though their report is not just a class project, but connects to a dynamic enterprise in the real world. Playing roles, even for short periods, can help students answer the "Who cares?" question, and help students see that projects of the sort they will do in the real world must have value to someone who is willing to pay for the work.

13.3 BUILD AN OUTLINE

Early, perhaps two weeks into the semester or quarter in high school or college, the instructor can ask each student to write an outline of their project and to bring it to the next class period. Specify the structure for the outline including a title, introduction, and other parts necessary to show heading and subheadings. The students should know the outline will not be graded, but simply give the instructor a chance to see student progress and provide feedback.

The instructor should collect the outlines, review them after class, and write comments on them to help students more fully develop their outline.

Begin with the title: is the title vague and general, or does the title tell the reader specifically what the project is about?

Look at the introduction: does the outline show development of background and rationale leading to specific objectives?

Continue through the outline providing comments, questions, and places where the author can further strengthen their writing and thinking. Return the outlines to each student, and allow

them to read the comments. Repeat the assignment to produce an outline on a regular basis until
the outlines show students are successfully moving forward.

13.4   EVALUATE SNACKS

Students in high school, college, and even in graduate school often fail to see the difference between descriptive and analytical writing. Part of the confusion comes when students assume that
writing that involves data must be analytical (which is not always the case). Instructors can use
snacks as an interesting teaching tool.

Count the number of students in the class and devise the snack activity based on student
pairs. If there are 30 students in the class, make 15 pairs. Give each student a small bag of chips and
a similar sized portion of convenience store pastry or cake. Each should cost about $1.

The activity is for each student pair to make a set of decisions about which snack they will
eat. The set of decisions might be cast as follows:

- **Decision 1.** Go ahead, each team make a choice based on "preference." Record the
  choices on the board.

- **Decision 2.** Each team becomes a "hypothetical person" with personality, preferences,
  tastes. The hypothetical person may be a vegetarian, favor diets low in salt, be on a low
  carbohydrate diet, be overweight, be underweight, be a teenager, be an adult, etc. After
  describing their "hypothetical person" the team then makes another choice. Answers go
  on the board.

  Have each team make a table that shows how their hypothetical person makes the
  snack choice based on the ingredients and information on the packaging. Have the
  teams identify relevant information, irrelevant information, and information gaps. What
  approach do they use for comparison: is a simple table adequate, or are there other approaches for decision making?

- **Decision 3.** Each team is the owner of a convenience store and wants to sell as many of
  these two snacks as possible. How does the storeowner decide how to arrange the store?

  One possibility is to put the snacks right at the front door, so customers see them when
  first entering. Another possibility is to put the chips snacks by the cool drinks, and to
  put the pastry snacks by the coffee. How would the storeowner make a decision about
  the placement? What experiment would help clarify the choice? How would the time
  of year and season affect customer choices in purchasing the snacks, and how would the
  owner test the effects of seasons on purchasing snacks?

The outcome of the snack exercise is that students begin to see how everyday choices involve some level of analysis, and that gathering information from food labels, and ideas for testing food placement in a store is more complex than first thought. Of course, students can also consume the snacks as they work through the discussions. Rather than using snacks, instructors can use many other kinds of products to get students to begin recognizing analytical thinking.

13.5   BRING A PIECE OF ANALYTICAL WRITING TO CLASS

To help students increase analytical content in their project with an activity scheduled just before they complete the first draft, instructors can give students the assignment to bring to the next class a hard copy of an analytical article related to their topic. During the next class, the instructor asks each student to spend a few minutes answering for themselves:

- What is the title?

- Who are the authors?

- Where do the authors work?

- What is analytical about the article?

After students have spent a few minutes thinking about the four questions, have a report out. Each student stands, and tells the class the title of the authors, who the authors are and where they work, and summarize why they feel the article is an example of analytical writing. Note the structure of the introduction, and specific objective(s). See the narrative explaining the approach, results, and significance of results.

After each student reports out, pass out an evaluation form, like the ones they will use to evaluate first drafts. Give the students 10 min of class period time to write an evaluation of the article they brought, to provide written comments about organization, content, format, and graphics, and to award the article points. Have another report out, where each student tells the class how many points the article earned, and why.

This exercise helps students understand the organization underlying analytical writing, and requires students have found at least one piece of analytical writing related to their topic. The exercise also gives students experience communicating orally in front of an audience, and reinforces the criteria for writing evaluation.

13.6   CREATE STORY PROBLEMS

Students often have trouble defining specific projects in their chosen subject areas. Such students often create outlines that will produce large, descriptive essays, or will have objectives that are far too large in scope to be resolved in the allotted time.

One approach to helping students create tractable goals is to spend a class period developing story problems. Story problems traditionally help students in K-12 learn arithmetic and how to operate fractions. For example, if a train goes 100 miles at 50 miles per hour, and then 200 miles at 100 miles per hour, how many total hours did the train travel? Of course, the answer is 4 h. Solving the problem requires three stages of analytical thinking.

1. Stating the specific objective. Determine the number of hours of train travel.

2. Developing the approach. Write a small equation explaining the logic behind solving the problem.

    Total time traveled = 100 mph/50 m + 200 mph/100 m

3. Finding the result = Solve the equation.

    Total time traveled = 2 h + 2 h = 4 h.

Students can create story problems to help develop appropriate goals and objectives for their analytical writing. One way to introduce story problems is to ask students to pair with a classmate. One student is to explain their theme area to another student, and to explain why the theme area is interesting and important. The other student listens, then creates a story problem for the first student to solve. The story problem must involve topics within the theme of interest. After some discussion between the students, a simple story problem will emerge. Now the student working on a project has an example of a specific objective, and the approach necessary to solve the problem.

Have each student pair explain to the class the project theme area and their story problem. Take time to explore the story problem for its analytical values, and the approach necessary to solve it. Now have the students in each pair reverse roles so by the end of class each student has a chance to explain their topic to another student, to take the lead in creating a story problem, and to have a story problem as an example of an objective they might use for their project.

Bibliography

Ammerman, J.W. 1993. Microbial cycling of inorganic and organic phosphorus in the water column. Pp. 649-660. In: Kemp, P.F., B.F. Sherr, E.B. Sherr, and J.J. Cole (eds.), *Handbook of Methods in Aquatic Microbial Ecology*, Lewis Publishers, Boca Raton, FL.

Black, R.A. and R.N. Mack. 1986. Mount St. Helens ash: recreating its effects on the steppe environment and ecophysiology. *Ecology* 67:1289-1302. DOI: 10.2307/1938685.

Brown, K. 2005. *Penicillin Man: Alexander Fleming and the Antibiotic Revolution.* The History Press. 368 p. 5

Brown, S. and A. E. Lugo. 1990. Tropical secondary forests. *Journal of Tropical Ecology* 6: 1-32. DOI: 10.1017/S0266467400003989. 62

Chemonics International, Inc. 2003. Community forestry management in the Maya Biosphere Reserve: Close to financial self-sufficiency? *Guatemala BIOFOR IQC Task Order 815.*

Costanza, R., R. d'Arge, R. de Groot, S. Farber, M. Grasso, B. Hannon, K. Limburg, S. Naeem, R. O'Neill, J Paruelo, R. Raskin, P. Sutton, and M. van den Belt. 1997. The value of the world's ecosystem services and natural capital. *Nature* 387: 253 – 260. 62

DOE. 1989. Cold Fusion Research. *A Report of the Energy Advisory Research Board to the United States Department of Energy.* DOE/S-0073 DE90 005611, 61 p. 86

Elias, M., and C. Potvin. 2003. Assessing inter- and intra-specific variation in trunk carbon concentration for 32 geotropically tree species. *Canadian Journal of Forest Resources* 33: 1039-1045. 62

Elkin, L. 2003. Rosalind Franklin and the double helix. Phys. *Today*, 56: 42-48. DOI: 10.1063/1.1570771. 4

Flemming, A. 1929. On the antibacterial action of cultures of Penicillium, with special reference to their use in the isolation of B. influenzae. *Brit. Jour. Exp. Path* (now *International Jour. Exp. Path*), 10:226-236. 6

Fleischmann, M. and S. Pons. 1989. Electrochemically induced nuclear fusion of deuterium. J. Electroanal. *Chem.* 261:3-1-308. 86

Gladstone, W.T., and F.T. Legid. 1990. Reducing pressure on natural forest through high-yield forestry. *Forestry Ecology and Management* 35: 69-78. 62

Guterman, L. 2006. A silent scientist under fire. *Chronicle Higher Ed.*, 52:15. 87

Houghton, R.A. 1996. Converting terrestrial ecosystems from sources to sinks of carbon. *Ambio* 25: 267-225. 62

Hwang, W.S., S.I. Roh, B.C. Lee, S.K. Kang, D.K. Kwon, S. Kim, S.J. Kim, S.W. Park, H.S. Kwon, C.K. Lee, J.B. Lee, J.M. Kim, C. Ahn, S.H. Paek, S.S. Chang, J.J. Koo, H.S. Yoon, J.H. Hwang, J.U. Hwang, Y.S. Park, S.K. Oh, H.S. Kim, J.H. Park, S.Y. Moon, and G. Schatten. 2005. Patient-specific embryonic stem cells derived from human SCNT blastocysts. *Science*, 308:1777-1783. DOI: 10.1126/science.1112286. 87

Keeling, C. D. and Whorf, T. P. 2004. Atmospheric Carbon Dioxide Concentrations at 10 Locations Spanning Latitudes 82°N to 90°S. ORNL/CDIAC-147, DOI: 10.3334/CDIA/atg.ndp001.2004, 10 p. 72

Larcher, W. 2003. *Physiological Plant Ecology: Ecophysiology and Stress Physiology of Functional Groups.* Springer, Berlin, Germany.

Madrid, S., and F. Chapela. 2003. Annex III: Certification in Mexico: The cases of Durango and Oaxaca. Pp. 1-2. In: Molnar, A. (Ed.). 2003. *Forest Certification and Communitieis: Looking forward to the Next Decade.* Pub. Forest Trends, Washington, D.C.

Mendel, G. 2008. *Experiments in Plant Hybridisation.* Cosimo Classics. 52 p. 3

Mendel, G. 1866. Versuche uber Pflanzen-Hybriden. Proceedings of the Natural History Society of Brunn (Verhandlungen des naturforschenden Vereins Brunn), 4:3-47 and in English 1901, J. R. Hortic. Soc. 26:1-32. 3

NOAA National Climatic Data Center, State of the Climate: Global Analysis for November 2012, published online December 2012, retrieved on December 17, 2012 from http://www.ncdc.noaa.gov/sotc/global/. 71

Nobel, P.S. 1991. *Physiochemical and environmental plant physiology.* Academic Press. 635 p.

Normile, D. 2009. Hwang convicted but dodges jail; stem cell research has moved on. *Science*, 326:650-651. DOI: 10.1126/science.326_650a. 88

Overton, J. 1994. *Sir Francis Bacon: A Biography.* George Mann Books. 384 p. 2

Parry M.L., O.F. Canziani, J.P. Palutikof, P.J. van der Linden and C.E. Hanson, Eds., 2007, *Climate Change 2007: Impacts, Adaptation and Vulnerability. Contribution of Working Group II to the Fourth Assessment Report of the Intergovernmental Panel on Climate Change,* Cambridge University Press, Cambridge, UK, 982pp. 93

Russell, M. 2013. *The Piltdown Man Hoax: Case Closed.* The History Press. 160 p. 85

Stern, M. 2004. As far as I can throw 'em: Expanding the paradigm for park/people studies beyond economic rationality. Paper presented at Conference, *People in Parks: Beyond the Debate.* Annual

Conference of the International Society of Tropical Foresters, Yale School of Forestry and Environmental Studies chapter, New Haven, Connecticut.

Tan, P. and R. Keeling. 2013. NOAA/ESLR (`www.eslr.noaa.gov/gmd/ccgg/trends`) and Scripps Institution of Oceanography (`scrippsco2.ucsd.edu/`). 69

Tans, P. 2013. NOAA/ESLR (`www.eslr.noaa.gov/gmd/ccgg/trends/`). 71

Taub, G. 1993. Bad Science: The Short Life and Weird Times of Cold Fusion. Random House, 503 p. 86

U.S. Surgeon General's Advisory Committee on Smoking and Health. 1964. Smoking and Health. Public Health Service. 387 p. 93

Watson, J.D. and F.H.C. Crick. 1953. A structure for Deoxyribose nucleic acid. Nature, 171:737-738. DOI: 10.1038/171737a0. 4

WRM (World Rainforest Movement). 2003. Ecuador: Mangroves and shrimp farming companies. http://www.wrm.org.uy/bulletin/51/Ecuador.html

Appendix

This handbook shows the linkages between analytical writing and thinking, and demonstrates the processes for outlining and writing the elements that constitute analytical writing. The ideas in this handbook apply to analytical writing in science, engineering, business, and any discipline or career founded on analysis of information. This handbook is not a technical writing guide, a writing style guide, or a writing guide for science and engineering. Here, students can find examples of writing guides to help with issues of technical writing, style, and issues specific to scientists and engineers. Additional resources important to handbook users are books on logic and reasoning.

TECHNICAL WRITING GUIDES

Technical writing guides assist those already working as, or studying to be, technical writers. Such guides instruct readers how to describe an engineered part, the steps for starting a machine or process, or a software function.

Lindsell-Roberts, S. 2001. *Technical Writing for Dummies.* For Dummies Pub. 336 p.

Van Laan, K., C. Julian, and J. Hackos. 2001. *The Complete Idiot's Guide to Technical Writing.* Alpha Pub. 352 p.

Pfeiffer, W. S. 2001. *Pocket Guide to Technical Writing.* Prentice Hall. 218 p.

WRITING STYLE GUIDES

Writing style guides help authors better use grammar, punctuation, spelling, voice and tense, and word choice.

Strunk, Jr., W. and E. B. White. M. Kalman (Illus.). 2005. *The Elements of Style Illustrated.* Penguin Press Pub. 176p.

Stillman, A. 1997. *Grammatically Correct: The Writer's Essential Guide to Punctuation, Spelling, Style, Usage and Grammar.* Writer's Digest Books Pub. 328 p.

Chicago Manual of Style Online. 2011. *The Chicago Manual of Style.* The University of Chicago Press Pub. 984 p.

SCIENCE AND ENGINEERING WRITING GUIDES

Most colleges and universities offer writing courses specifically for science and engineering students, and use either existing volumes or online resources as instructional aids.

The vast majority of science and engineering students do not take advanced writing courses, but still must complete a capstone course that requires a written project. Many science and engineering writing guides are not standalone teaching tools, or useful for high school students, students in community colleges, or those in professional careers. Few science and engineering writing guides address the connections between analytical writing and strategic thinking. Instead, these guides focus largely on matters of writing formats and conventions used in science.

Alley, M. 1996. *The Craft of Scientific Writing*. Springer. 289 p. The book aims to promote effective scientific communication for scientists and engineers. Nine of the 14 chapters focus on language skills to increase precision, clarity, and concise writing. The first chapter explains "Where to Begin" and the last chapter explains "Actually Sitting Down to Write."

Goldbort, R. 2006. *Writen for Science*. Yale University Press. 330 p. The book is a comprehensive volume of 10 chapters, each dedicated to a specific type of scientific writing product. Each chapter focuses on a single writing product such as laboratory notes, dissertations, graduate papers, journal articles, and grant proposals.

Porush, D. 1995. *A Short Guide to Writing about Science*. HarperCollins. 275 p. The book is one of the Short Guide to Writing Series that also includes books on writing for science, biology, social sciences, chemistry, art, history, and music. *The Short Guide to Writing about Science* is a popular volume for science writing classes at colleges and universities, and explains how to write laboratory reports, research papers, and descriptive essays in science. The approach takes students through 12 chapters that systematically focus on the mechanics of collecting data, writing a title, abstract, introduction, methods, results, testing hypotheses, and references.

Rosenwasser, D. and J. Stephen. 2011. *Writing Analytically*. Thomson Wadsworth. 416 p. The book is a textbook used for teaching analytical writing to college and university students. The book does more than is necessary for students and teachers. For example, there is substantial discussion of revising style, writing about reading, and a conversation model for using sources.

BOOKS ON LOGIC AND REASONING

Logic and reasoning comprise a formal area of study more than 2,000 years old.

Kelley, D. 1998. *The Art of Reasoning*. W. W. Norton and Company. 704 p. Instructors often use this book as an introduction to reasoning and logic. The author includes both critical reasoning and Aristotelian logic and introduces students to deductive logic, classification of information and ideas, definitions of terms used in logic and reasoning, and analysis of arguments.

Hausman, A., J. Kahane, and P. Tidman. *Logic and Philosophy: A Modern Introduction*. Wadsworth. 434 p. The book is a comprehensive and rigorous introduction to the theories, concepts, and definitions of logic. The book uses truth tables and reasoning underlying proofs.

About the Author

Dr. William E. Winner is a leader in developing university programs in the areas of energy, environment, and sustainability. He earned B.A, M.A.T., M.A., and Ph.D. degrees, and held academic positions, at Virginia Tech and Oregon State University prior to his appointment to North Carolina State University. Dr. Winner also worked in research administration at the National Science Foundation. At NC State University, Dr. Winner is Director of the Environmental Sciences Academic Program, chairs the University Energy Council, and is Co-Chair of the Campus Environmental Sustainability Team. He serves on advisory boards for EnvironMentors, the Institute for Sustainable Development, and the Klamath Basin Rangeland Trust. He represents NC State University in the National Council for Science and the Environment, and other national organizations including the Council of Environmental Deans and Directors, and the Council for Energy Research and Education Leaders. Dr. Winner has published and reviewed numerous articles for peer reviewed journals and co-edited two books. He reviews proposals for competitive grants and scholarships and editorial work for two journals.